01-ECO-1

Scientific Review of Cyanide Ecotoxicology and Evaluation of Ambient Water Quality Criteria

by:

Robert W. Gensemer, Ph.D.
David K. DeForest
Rick D. Cardwell, Ph.D.
Parametrix, Inc.

David Dzombak
Carnegie Mellon University

Robert Santore
HydroQual, Inc.

2007

WERF

The Water Environment Research Foundation, a not-for-profit organization, funds and manages water quality research for its subscribers through a diverse public-private partnership between municipal utilities, corporations, academia, industry, and the federal government. WERF subscribers include municipal and regional water and wastewater utilities, industrial corporations, environmental engineering firms, and others that share a commitment to cost-effective water quality solutions. WERF is dedicated to advancing science and technology addressing water quality issues as they impact water resources, the atmosphere, the lands, and quality of life.

For more information, contact:
Water Environment Research Foundation
635 Slaters Lane, Suite 300
Alexandria, VA 22314-1177
Tel: (703) 684-2470
Fax: (703) 299-0742
www.werf.org
werf@werf.org

This report was co-published by the following organizations. For non-subscriber sales information, contact:

IWA Publishing
Alliance House, 12 Caxton Street
London SW1H 0QS, United Kingdom
Tel: +44 (0) 20 7654 5500
Fax: +44 (0) 20 7654 5555
www.iwapublishing.com
publications@iwap.co.uk

Library of Congress Catalog Card Number: 2005938575
IWAP ISBN: 978-1-84339-753-3

Parametrix, Inc.

ACKNOWLEDGMENTS

The research team thanks Dr. Richard Caldwell and his team at Northwestern Aquatic Sciences for conducting the *Cancer irroratus* toxicity tests. They also thank Dr. Eric Van Genderen, Allison Cardwell, and Carrie Smith at the Parametrix Environmental Research Laboratory (PERL) for conducting the *Acartia tonsa* toxicity tests; organisms were provided by AlgaGen, Inc. Amphibian toxicity tests were conducted by ENSR International with assistance from Dr. David Pillard and Dr. Rami Naddy. Additional technical assistance was provided by Jeremy Clark, Sharon Grimm, Cortney Higgins, Angela Stenhouse, and Dr. William Stubblefield. This research was funded by the Water Environment Research Foundation.

Report Preparation

Principal Investigator:
Robert W. Gensemer, Ph.D.
Parametrix, Inc.

Project Team:
Rick D. Cardwell, Ph.D.
Parametrix, Inc.

David K. DeForest
Parametrix, Inc.

David A. Dzombak, Ph.D.
Carnegie Mellon University

Robert C. Santore
HydroQual, Inc.

Project Subcommittee

Walter Berry, Ph.D.
U.S. Environmental Protection Agency

Philip B. Dorn, Ph.D.
Shell Global Solutions U.S., Inc.

Joseph W. Gorsuch
Gorsuch Environmental Management Services, Inc.

James J. Pletl, Ph.D.
Hampton Roads Sanitation District

Mary C. Reiley
U.S. Environmental Protection Agency

Water Environment Research Foundation Staff

Director of Research: Daniel M. Woltering, Ph.D.
Program Director: Margaret Stewart

ABSTRACT AND BENEFITS

Abstract:

Current national ambient water quality criteria (AWQC) for cyanide were developed in 1984. This study reviewed and updated our current scientific knowledge on the aquatic chemistry and toxicity of cyanide and new aquatic toxicity studies were conducted to fill critical data gaps. Recalculated acute and chronic freshwater criteria are 23 and 4.8 μg CN/L, respectively, which are similar to the current National acute and chronic criteria of 22 and 5.2 μg CN/L. The recalculated acute marine criterion of 5.5 μg CN/L is 5-fold higher than the current acute marine criterion of 1.0 μg CN/L, largely due to additional toxicity testing with the East Coast rock crab (*Cancer irroratus*) and other *Cancer* species. The recalculated chronic marine criterion of 1.1 μg CN/L is only slightly different than the current chronic marine criterion of 1.0 μg CN/L due to changes in the recalculated acute-chronic ratio used to derive the chronic criterion from acute toxicity data. Consistent with the U.S. EPA's draft strategy for developing future AWQC, recalculated freshwater criteria also would likely protect aquatic-dependent wildlife, benthic organisms, and most freshwater threatened and endangered species. Finally, cyanide AWQC should be implemented on the basis of "free" rather than "total" cyanide chemical forms.

Benefits:

- Uses the state-of-the-science on cyanide chemistry and aquatic toxicology to suggest recalculations of AWQC that are scientifically defensible and protective of aquatic life and their uses in our nation's waters
- Resolves key scientific uncertainties in the acute toxicity of cyanide to marine organisms, including rock crabs (genus *Cancer*) and marine copepods (genus *Acartia*)
- Supports regulatory evaluations of AWQC for protection of threatened and endangered species, aquatic-dependent wildlife, and benthic organisms
- Provides scientific support for regulating cyanide discharges on the basis of analytical measurements of free, rather than total, cyanide as is currently recommended in the National AWQC.

Keywords: Cyanide, toxicity, threatened and endangered species, wildlife, benthos.

TABLE OF CONTENTS

LIST OF TABLES

Text Tables:

Appendix: Data Tables to Support Criteria Calculations:

LIST OF FIGURES

LIST OF ACRONYMS AND ABBREVIATIONS

ACR	Acute-Chronic Ratio
Ag	Silver
Al	Aluminum
ASTM	American Society for Testing and Materials
AWQC	Ambient Water Quality Criteria
B	Boron
Ba	Barium
Be	Beryllium
Br	Bromide
BW	Body Weight
Ca	Calcium
CCC	Criterion Continuous Concentration
Cd	Cadmium
Cl	Chloride
CMC	Criterion Maximum Concentration
CN	Cyanide
Cr	Chromium
C_s	Chemical Sediment Concentration
Cu	Copper
C_w	Chemical Water Concentration
CWA	Clean Water Act
EC20	20% Effect Concentration
EC50	50% Effect Concentration
EqP	Equilibrium Partitioning
ESA	Endangered Species Act
F	Fluoride
FAV	Final Acute Value
FCV	Final Chronic Value
Fe	Iron
f_{oc}	Fraction Organic Carbon
Free Cyanide	HCN and CN^-
FWS	Fish and Wildlife Service
GMAV	Genus Mean Acute Value
g_s	Grams of Sediment
H_3BO_3	Boric Acid
HCN	Hydrogen Cyanide
HCO_3	Bicarbonate
ICE	Interspecies Correlation Estimations Model
K	Potassium
KCN	Potassium Cyanide
K_{oc}	Organic Carbon-Normalized Sediment Partition Coefficient
LC01	Lethal Concentration to 1% of Test Organisms
LC50	Lethal Concentration to 50% of Test Organisms
LD50	Lethal Dose to 50% of Test Organisms

LOAEL	Lowest Observed Adverse Effect Level
LOEC	Lowest Observed Effect Concentration
Mg	Magnesium
Mn	Manganese
Mo	Molybdenum
MOA	Memorandum of Agreement
MST	Median Survival Time
Na	Sodium
NaCN	Sodium Cyanide
NaOH	Sodium Hydroxide
NH_3	Ammonia
Ni	Nickel
NMFS	National Marine Fisheries Service
NOAEL	No Observed Adverse Effect Level
NOEC	No Observed Effect Concentration
P	Phosphorus
Pb	Lead
pK_a	Log Acid Dissociation Constant
PO_4	Phosphate
Si	Silica
SMAV	Species Mean Acute Value
SO_4	Sulfate
Sr	Strontium
SSD	Species Sensitivity Distribution
T&E	Threatened and Endangered
TRAP	Toxicity Relationship Analysis Program
U.S. EPA	United States Environmental Protection Agency
UV	Ultraviolet
WAD	Weak Acid Dissociable Cyanide
WERF	Water Environment Research Foundation
WIR	Water Ingestion Rate
WWTP	Wastewater Treatment Plant
Zn	Zinc

Executive Summary

Background

The U.S. EPA's current ambient water quality criteria (AWQC) for cyanide were developed in 1984. However, since 1984 new scientific information on cyanide toxicity and chemistry has been published, which suggested that the existing cyanide criteria should be reassessed. Accordingly, WERF funded a study to evaluate the cyanide chemistry and aquatic toxicity data published over the last two decades and the potential influence of these data on cyanide criteria development and implementation. Traditionally, the U.S. EPA's AWQC for protection of aquatic life are based on the sensitivity of aquatic biota, such as fish, invertebrates, aquatic plants and algae, to chemicals from water-based exposures. However, in response to the U.S. EPA's draft strategy for developing future AWQC, this study also considers whether criteria are protective of benthic (sediment) organisms, threatened and endangered (T&E) species, and aquatic-dependent wildlife.

Research Objectives

The most up-to-date research on cyanide chemistry and aquatic toxicity from the scientific literature was reviewed, and new cyanide toxicity studies were conducted to fill critical data gaps in our understanding of the cyanide concentrations and chemical forms of cyanide required to protect aquatic organisms. The ultimate goal was to use the methods developed by the U.S. EPA to derive AWQC as a format for evaluating the scientific basis for regulating cyanide discharges to fresh and marine receiving waters. It must be noted that although this evaluation was conducted collaboratively with the U.S. EPA, the results of the study have not been officially reviewed and approved by the U.S. EPA. The specific research objectives included the following:

- Review the scientific literature published since the 1984 cyanide AWQC were developed to identify relevant new information on cyanide chemistry and toxicity;
- Conduct new toxicity tests to fill critical data gaps in our understanding of cyanide chemistry and toxicity;
- Evaluate cyanide chemistry and bioavailability to determine what chemical form of cyanide (and its measurement) should be used to implement cyanide AWQC;
- Synthesize the new scientific information by recalculating cyanide AWQC according to U.S. EPA guidance; and
- Evaluate whether recalculated cyanide criteria are protective of benthic organisms, T&E species, and aquatic-dependent wildlife.

Ambient Water Quality Criteria

The U.S. EPA's AWQC set maximum concentrations of chemicals intended to be protective of aquatic life in a wide range of freshwater and marine environments. These criteria are derived from laboratory toxicity data and are generally calculated based on the 5th percentile sensitivity of tested organisms. However, a calculated value may be lowered if it appears to be under-protective of commercially or recreationally important species (e.g., rainbow trout).

Briefly, the first step in developing AWQC is to compile acute and chronic toxicity data that meet U.S. EPA guidelines for criteria derivation. For each species with acceptable acute toxicity data, the species mean acute value (SMAV) is calculated as the geometric mean of available 48- to 96-hr LC/EC50s (i.e., concentrations at which acute impacts such as mortality occur in 50% of the tested organisms) for each species. The genus mean acute value (GMAV) is then calculated as the geometric mean of available SMAVs for each genus. The 5th percentile of the distribution of available GMAVs is identified as the final acute value (FAV), which is divided by two to determine the criterion maximum concentration (CMC) which is more commonly termed the "acute criterion." It is important to note that the 5th percentile is calculated based solely on the four most sensitive GMAVs and the total number of GMAVs.

The chronic criterion may be derived in a manner similar to the CMC, but chronic toxicity data are typically unavailable for a sufficient number of species (as is the case for cyanide). It is thus typically necessary to apply an acute-chronic ratio (ACR) to the FAV to estimate the final chronic value (FCV). Unless other data are available to suggest the FCV is under-protective of the aquatic community (including aquatic plants and protection from bioaccumulative substances), the criterion continuous concentration (CCC), or "chronic criterion," is set equal to the FCV.

New Scientific Studies

New scientific studies with cyanide were evaluated from 1) papers published over the last two decades, and 2) new original scientific research:

- Freshwater Organisms: New toxicity studies that meet U.S. EPA guidance for AWQC development were identified for 13 freshwater genera, including new toxicity studies conducted with three species of larval amphibians (*Rana pipiens*, *Rana berlandieri*, *Xenopus laevis*) and the threespine stickleback (*Gasterosteus aculeatus*).
- Saltwater Organisms: New toxicity studies that meet U.S. EPA guidance for AWQC development were identified for just one aquatic genus, but five additional species (mostly within the genus *Cancer*). In addition, new toxicity studies were conducted with the acutely sensitive rock crab Cancer irroratus and marine copepod (*Acartia tonsa*).
- Bioavailability and Measurement of Cyanide Toxicity: Studies confirmed that free cyanide (HCN and CN-) are the most toxic to aquatic organisms and because standard methods now exist for the measurement of free cyanide (ASTM D4282-95), this would be the most accurate form for measuring compliance with AWQC or standards.

Recalculated Cyanide AWQC

Based on the new toxicity information summarized above, recalculated acute and chronic freshwater criteria for cyanide are 23 and 4.8 μg CN/L, respectively, which are similar to the current U.S. EPA acute and chronic criteria of 22 and 5.2 μg CN/L. Because salmonid fishes are uniformly the most sensitive to acute cyanide exposures, the researchers also recalculated acute and chronic criteria to be 47 and 9.8 μg CN/L, respectively, for use in waters without resident salmonids (e.g., warm waters). The recalculated acute marine criterion of 5.5 μg CN/L is 5-fold higher than the current U.S. EPA acute marine criterion of 1.0 μg CN/L, largely due to additional toxicity testing with the East Coast rock crab (*Cancer irroratus*) and other Cancer species. The recalculated chronic marine criterion of 1.1 µg CN/L is only slightly different than the current U.S. EPA chronic marine criterion of 1.0 µg CN/L due to changes in the recalculated acute-chronic ratio used to derive the chronic criterion from acute toxicity data. A summary of the recalculated freshwater and marine criteria relative to the current U.S. EPA freshwater and marine criteria are provided in Tables ES-1 and ES-2, respectively.

Table ES-1. Comparison of 1984 National Criteria with Recalculated Acute and Chronic Cyanide Criteria for Freshwater Organisms.

	1984 AWQC		Recalculated: All Species		Recalculated: Excluding Salmonids	
GMAV Rank	Genus	GMAV (µg CN/L)	Genus	GMAV (µg CN/L)	Genus	GMAV (µg CN/L)
4	*Lepomis*	99.28	*Perca*	92.64	*Pomoxis*	102.0
3	*Perca*	92.64	*Salmo*	90.00	*Lepomis*	99.28
2	*Salvelinus*	85.8	*Salvelinus*	85.80	*Gasterosteus*	98.80
1	*Salmo*[a]	63.45	*Oncorhynchus*[a]	46.53	*Perca*	92.64
Final Acute Value (FAV)		44.73[b]		46.53[b]		94.50
Acute Criterion (CMC)		***22***		***23***		***47***
Acute to Chronic Ratio (ACR)		8.568		9.659		9.659
Chronic Criterion (CCC)		***5.2***		***4.8***		***9.8***

[a] Rainbow trout were classified under the genus Salmo in 1984, but have since been reclassified under the genus *Oncorhynchus*
[b] Lowered from calculated value to protect rainbow trout
GMAV = Genus mean acute value
Note: all concentrations are in µg/L free cyanide expressed as CN

Table ES-2. Comparison of 1984 National Criteria with Recalculated Acute and Chronic Cyanide Criteria for Marine Organisms.

	1984 AWQC		Recalculated	
GMAV Rank	Genus	GMAV (µg CN/L)	Genus	GMAV (µg CN/L)
4	*Mysidopsis*[a]	118.4	*Americamysis*[a]	118.4
3	*Menidia*	59	*Cancer*	84.69
2	*Acartia*	30	*Menidia*	59
1	*Cancer*	4.893	*Acartia*	17
Final Acute Value (FAV)		2.030		11.00
Acute Criterion (CMC)		***1.0***		***5.5***
Acute to Chronic Ratio (ACR)		2		9.659
Chronic Criterion (CCC)		***1.0***		***1.1***

[a] The genus *Mysidopsis* has been changed to *Americamysis* since 1984
GMAV = Genus mean acute value
Note: all concentrations are in µg/L free cyanide expressed as CN

Limited empirical cyanide toxicity data are available for freshwater T&E species, but a larger data set of modeled toxicity data for T&E species was evaluated using U.S. EPA's Interspecies Correlation Estimations (ICE) model. Based on the conservative nature of this T&E analysis, it appears that the recalculated criteria would be protective of most T&E species (possible exceptions based on ICE predictions include the fountain darter, *Etheostoma fonticola*, and Apache trout, *Oncorhynchus mykiss*). Similarly, recalculated freshwater cyanide AWQC should be protective of wildlife ingesting surface water because these concentrations are below those that would be expected to pose a risk to wildlife species based on consideration of avian and mammalian toxicity thresholds and representative surface water ingestion rates for birds and mammals. Finally, recalculated AWQC also should be protective of benthic organisms because 1) invertebrates tend to be less sensitive than the fish that drive the basis for the recalculated criteria; 2) cyanide does not bind strongly to most aquatic sediments; and 3) cyanide is rapidly metabolized by many organisms and is thus unlikely to persist in freshwater sediments.

CHAPTER 1.0

INTRODUCTION

Ambient water quality criteria (AWQC) for cyanide were developed in 1984 (U.S. EPA, 1985a). Recently, however, concerns have arisen that the AWQC for cyanide have been problematic to implement and may not accurately reflect either the toxic forms or bioavailable concentrations of cyanide in water, sediments, and tissues of aquatic organisms. Also, cyanide criteria typically have been implemented based on total cyanide concentrations rather than the free cyanide concentrations that formed the basis of the criteria calculations (U.S. EPA, 1985a, 2002). New knowledge on cyanide toxicity, cyanide speciation and its measurement, and the relative toxicity of bioavailable cyanide species suggested that a re-evaluation of the aquatic toxicity data and chemistry that serve as the basis of the current national criteria is warranted.

In response to these concerns, studies sponsored by the Water Environment Research Foundation (WERF) evaluated the formation and fate of cyanide in municipal wastewater effluents (Kavanaugh et al., 2003). Kavanaugh et al. evaluated the reliability of several analytical methods for measurement of cyanide species at low (ppb) concentrations, along with studies to evaluate the sources, transport, and fate of cyanide species in wastewater treatment plants (WWTPs) and receiving waters. Results indicated that potentially toxic forms of cyanide can be generated in WWTPs from several chemical processes, including dissociation of thiocyanate by chlorination or UV disinfection, chlorination in the presence of residual ammonia, nitrosation, and photolysis of ferrocyanate. Models were developed for predicting the fate of these compounds in surface waters downstream of WWTP effluents to establish where organisms might be most at risk from cyanide exposure. While implementation of these results could assist in the development of site-specific water quality standards for cyanide, a more thorough update to the National AWQC was warranted to apply our more current scientific understanding of cyanide toxicology to the derivation of aquatic life criteria.

Therefore, the present study was undertaken to review and expand the aquatic toxicity database for cyanide, and to evaluate how this updated cyanide toxicity database could influence the recalculation of the national AWQC and thereby enhance the scientific basis of the concentrations and chemical forms of cyanide specified for protection of aquatic organisms. This review and AWQC recalculation was conducted according to U.S. EPA guidance for derivation of AWQC for protection of aquatic life (U.S. EPA, 1985b), and is based both on an extensive literature review (Clark et al., 2006; Gensemer et al., 2006), and on new toxicity studies that fill key data gaps. Also, consistent with the U.S. EPA's *Draft Strategy: Proposed Revisions to the "Guidelines for Deriving Numerical National Water Quality Criteria for the Protection of Aquatic Organisms and Their Uses"*(U.S. EPA, 2003b), this re-evaluation of cyanide criteria includes consideration of impacts on benthic (sediment) organisms, wildlife that consume aquatic biota, and possible effects to threatened and endangered (T&E) species to ensure that revised criteria are adequately protective of the entire aquatic ecosystem. It must be emphasized that even though WERF has worked collaboratively with the U.S. EPA on this project, these recalculated cyanide criteria have not been officially reviewed by the U.S. EPA. Therefore, this report does not necessarily reflect the U.S. EPA's position with regard to AWQC for cyanide.

1.1 National AWQC

National AWQC for protection of aquatic life set maximum threshold concentrations of contaminants for both freshwater and marine environments. These criteria are derived from empirical toxicity data and are designed to be stringent enough to protect most sensitive species potentially exposed to a contaminant in any waterbody in the United States. Below these thresholds, no adverse effects on aquatic community function are anticipated, although the most sensitive species could be impacted as the AWQC are designed to protect all but the most sensitive 5% of species. If data suggest that a commercially or recreationally important species is not protected at this level, then an AWQC value can be adjusted to provide sufficient protection for these species as well.

To understand how AWQC are developed, it is useful to review the guidelines and terminology provided in U.S. EPA (1985b), but the general approach is briefly summarized below. The first step is to compile acute and chronic toxicity data that meet the U.S. EPA (1985b) guidelines. For each species with acceptable acute toxicity data, the species mean acute value (SMAV) is calculated as the geometric mean of available 48 to 96-hr LC/EC50s for each species. The genus mean acute value (GMAV) is then calculated as the geometric mean of available SMAVs for each genus. The 5th percentile of the distribution of available GMAVs is identified as the final acute value (FAV), which is divided by two to determine the criterion maximum concentration (CMC) which is more commonly termed the "acute criterion." It is important to note that the 5th percentile is calculated based solely on the four most sensitive GMAVs and the total number of GMAVs (U.S. EPA, 1985b).

The chronic criterion may be derived in a manner similar to the CMC, but chronic toxicity data are typically unavailable for a sufficient number of species (as is the case for cyanide). It is thus typically necessary to apply an acute-chronic ratio (ACR) to the FAV to estimate the final chronic value (FCV). Unless other data are available to suggest the FCV is under-protective of the aquatic community (including aquatic plants and protection from bioaccumulative substances), the criterion continuous concentration (CCC), or chronic criterion, is set equal to the FCV.

1.2 Study Goals and Report Layout

The overall objectives of the project were to evaluate existing knowledge of cyanide aquatic toxicity, identify data gaps, and fill these data gaps to support recalculations of AWQC for the protection of aquatic life from cyanide (including aquatic-dependent wildlife and benthic organisms). This project was conducted in two phases: Phase 1 consisted of a database and literature review for cyanide chemistry (including fate, speciation, and analytical methods), and toxicity to aquatic biota (both for pelagic and benthic organisms) and aquatic-dependent wildlife. From these reviews, recommendations were made to conduct new toxicity testing, which consisted primarily of acute cyanide toxicity tests to the East Coast rock crab, *Cancer irroratus*, and the marine copepod, *Acartia tonsa*[1]. The second phase of the project consisted of publishing the literature reviews in the primary scientific literature, and generation of this final report which provides recalculations of the national AWQC for cyanide.

This report is written according to U.S. EPA guidance for preparation of AWQC for protection of aquatic life (U.S. EPA, 1985b), with modifications made to accommodate discussions of AWQC protection for T&E species, aquatic-dependent wildlife, and benthic organisms. Section 2 of the report describes cyanide sources and chemistry, discusses analytical methods for measuring free cyanide, and implications for interpreting levels of cyanide that are toxic to aquatic life. Sections 3 through 6 and 10 through 11 are consistent with the traditional organization of AWQC documents and present aquatic toxicity and bioaccumulation data for aquatic life. Sections 7 through 9 present special cyanide studies related to cyanide toxicity to freshwater T&E species, aquatic-dependent wildlife, and benthic organisms. The final section of the report summarizes the key issues from the previous sections and provides recalculations of cyanide AWQC.

---

[1] Previous cyanide testing was conducted using *Acartia clausi*, but the limited commercial availability of this species necessitated the use of *Acartia tonsa*.

CHAPTER 2.0

CYANIDE CHEMISTRY AND TOXICITY STUDY REVIEW

The cyanide group (CN) can be present in a number of different compounds, such as free cyanide (HCN and CN^-), simple cyanide salts, (e.g. KCN, NaCN), metal-cyanide complexes, and in some organic compounds (Higgins and Dzombak, 2006). Free cyanide is the sum of the hydrocyanic acid, HCN, and the cyanide ion, CN^-. The total cyanide measurement captures free cyanide plus metal-cyanide complexes. The distribution of free cyanide between HCN and CN^- depends on pH and temperature. The pK_a for hydrocyanic acid is 9.23 at 25°C, indicating that for water below a pH of 9 the predominant form of free cyanide is HCN, which is volatile.

The cyanide ion, CN^-, is a versatile and sometimes strong complexing agent. Thus, cyanide often occurs in predominantly metal-complexed forms in many natural waters. Simple cyanide salts, such as NaCN and KCN, are highly soluble and dissociate completely in ambient water. Metal-cyanide complexes such as zinc-, nickel-, copper-, and cadmium-cyanide easily dissociate under acidic conditions and are classified as weak acid dissociable (WAD) complexes (American Public Health Association, 1998; Zheng et al., 2003). Strong complexes do not dissociate easily and include cobalt- and iron-cyanide complexes. For any metal-cyanide complex, the rate of dissociation is governed by solution conditions, such as water temperature, pH, major ion concentrations, and the concentration of the complex. Simple metal-cyanide complexes dissociate readily under most conditions in aqueous systems. Complex cyanide species, such as iron-cyanide species, are stable in the absence of light, but can dissociate readily when light is present (Meeussen et al., 1992). The rate of dissociation is also dependent on pH.

As discussed in the AWQC for cyanide (U.S. EPA, 1985a), the apparent toxicity of most simple cyanides and metallocyanide complexes to aquatic life mainly results from HCN derived from dissociation, photodecomposition, and hydrolysis (Doudoroff et al., 1966; Smith et al., 1979; Pablo et al., 1997a; Pablo et al., 1997c), although CN^- is apparently also toxic (Broderius et al., 1977). Although there are suggestions in the available literature that some metal-cyanide complexes may be as toxic as free cyanide, a more detailed analysis of these data suggests that the toxicity of these complexes is actually due to the associated increase in metal ion concentration (Section 2-1). Overall, free cyanide (HCN and CN^-) thus appears to be the most toxicologically relevant form to aquatic life. Indeed, the 1984 AWQC for cyanide are specified in terms of free cyanide (U.S. EPA, 1985a). For a more detailed review of the toxicological importance of free cyanide to aquatic organisms, see Gensemer et al. (2006).

As discussed in the sections that follow, understanding the chemical fate of metal-cyanide complexes, their relative bioavailabilities to aquatic organisms, and properly measuring the toxic cyanide species are critical for developing and implementing appropriate AWQC for cyanide.

2.1 Bioavailability of Metal-Cyanide Complexes

Although there is a general consensus in the literature that metal-cyanide complexes are much less toxic than free cyanide, there have been suggestions by some authors that some metal-cyanide complexes may be as toxic as free cyanide. Notably, Broderius (1973) reported observed toxicity when fish were exposed to varying concentrations of Ni-CN species and free cyanide. The composition of the mixtures was designed to provide a range of total cyanide concentrations, while maintaining a nearly constant concentration of free cyanide. It was expected, therefore, that in these mixtures the constant free cyanide would result in nearly constant toxicity. Contrary to expectations, the toxicity varied such that mixtures with higher total cyanide (in the form of metal-cyanide complexes) were more toxic (Figure 2-1), and from these observations Broderius concluded that Ni-CN complexes (and specifically the $Ni(CN)_4^{-2}$ ion) were toxic to fish (Broderius, 1973).

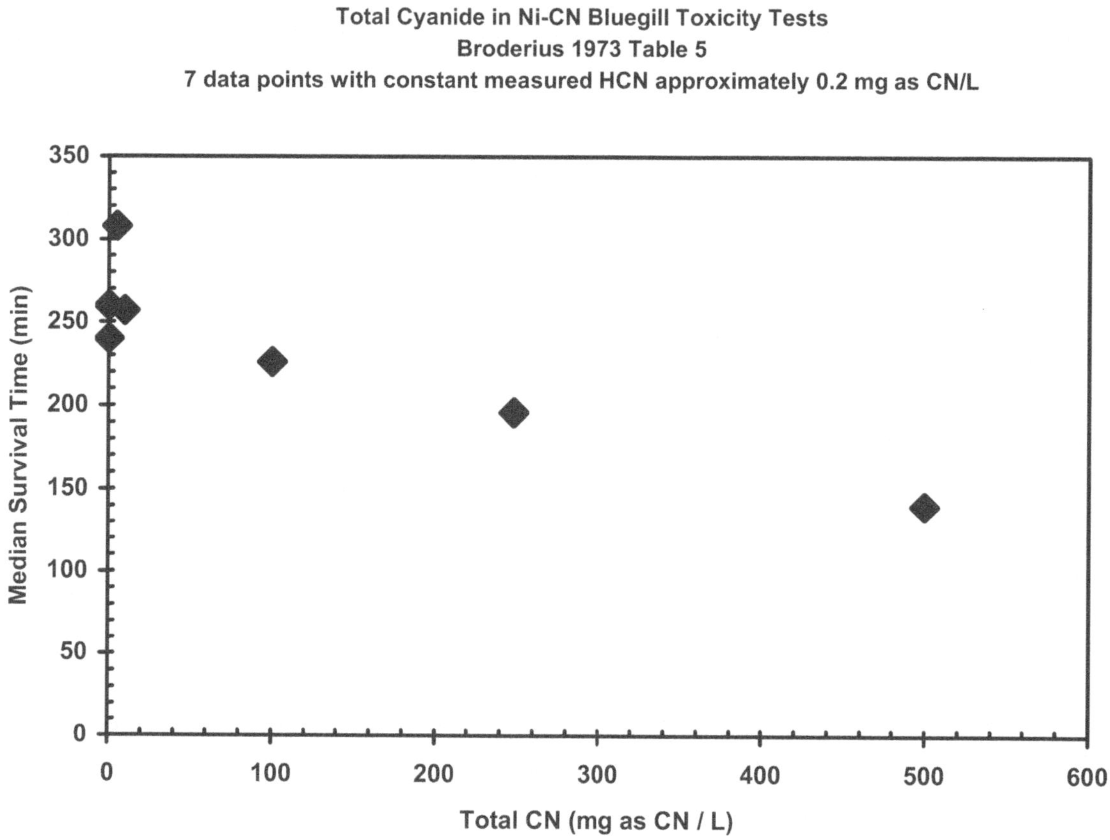

Figure 2-1. Median Survival Time of Bluegills In the Presence of Cyanide-Nickel Mixtures Designed So That Free Cyanide Was Constant.

In another set of experiments Broderius noted the potential toxicity of Ag-CN complexes. In these experiments, it was observed that median survival time (MST) of threespine sticklebacks (*Gasterosteus aculeatus*) was related to free cyanide concentration (triangles and solid trendline in Figure 2-2), but lower MSTs were noted when Ag-CN was the source of the free cyanide (diamonds and dashed trendline in Figure 2-2), suggesting some additional toxicity resulting from Ag-CN (i.e., $Ag(CN)^{2-}$) beyond that attributable to free cyanide. Since these conclusions are incompatible with the central tenet that HCN governs cyanide toxicity, they warrant further scrutiny.

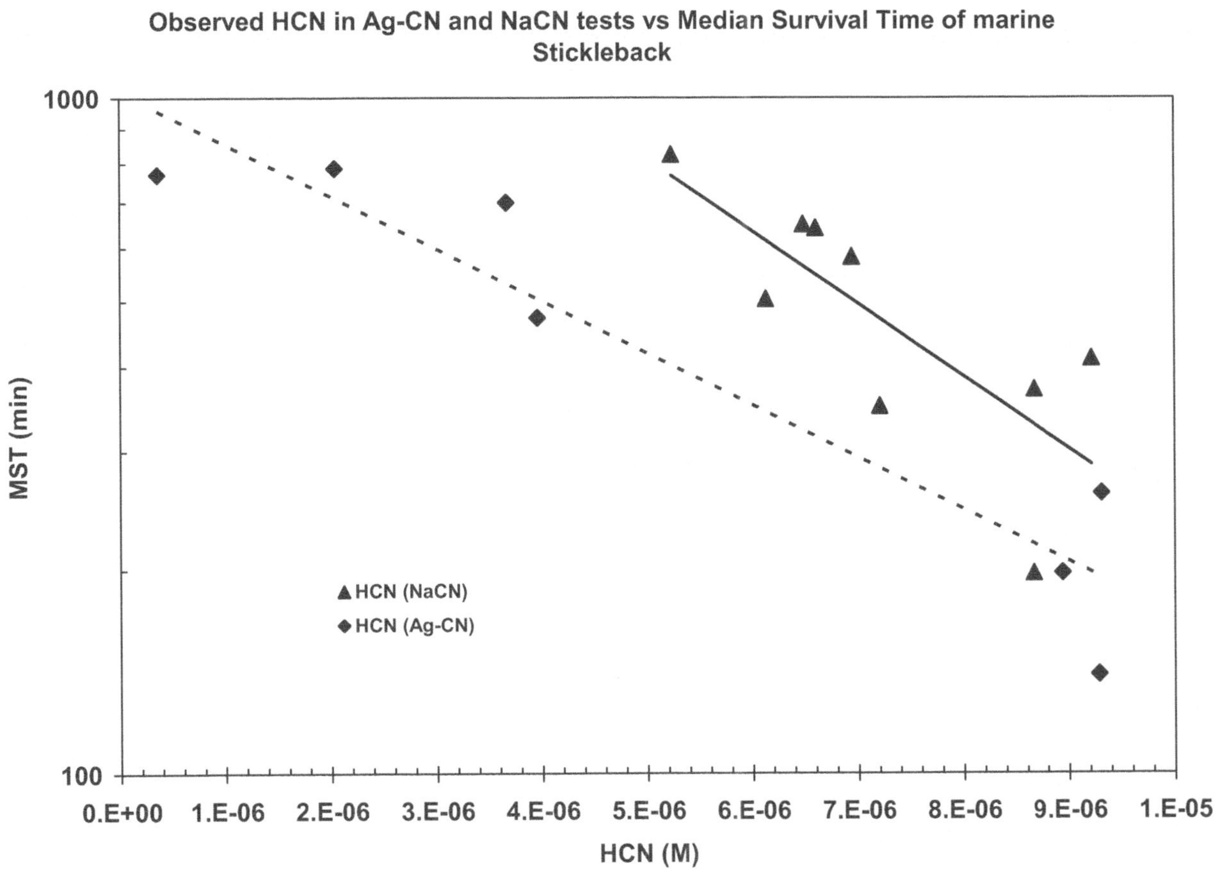

Figure 2-2. Median Survival Time of Threespine Sticklebacks (*G. Aculeatus*) In the Presence of Cyanide-Silver Mixtures.

The chemical conditions used by Broderius (1973) in the Ni-CN and Ag-CN toxicity tests were examined with a chemical equilibrium model to estimate the concentrations of individual species in each of the toxicity exposures. The simulated chemical reactions included metal-cyanide complexation as well as other metal-complexation reactions needed to determine metal speciation (Gensemer et al., 2003) and was solved using the CHESS model (Santore and Driscoll, 1995).

While it is true that toxicity increased as total cyanide increased (even though free cyanide was the same in these exposures) the total cyanide concentrations were adjusted by adding a Ni-CN salt (Figure 2-1). As a result, the total nickel concentration was also changing. Chemical speciation calculations show that there was sufficient nickel in these exposures to potentially cause toxicity, and that the nickel present was in a bioavailable form. Free nickel ion concentrations were as high as free nickel ion concentrations calculated at the LC50 values for some other aquatic organisms using the Biotic Ligand Model (Figure 2-3) (Di Toro et al., 2001; Santore et al., 2001). Furthermore, the exposures with the highest free nickel ion concentration in Figure 2-3 are the exposures with the shortest MST. It is probable, therefore, that the toxicity observed by Broderius was caused by the nickel ion and not by nickel-cyanide complexes.

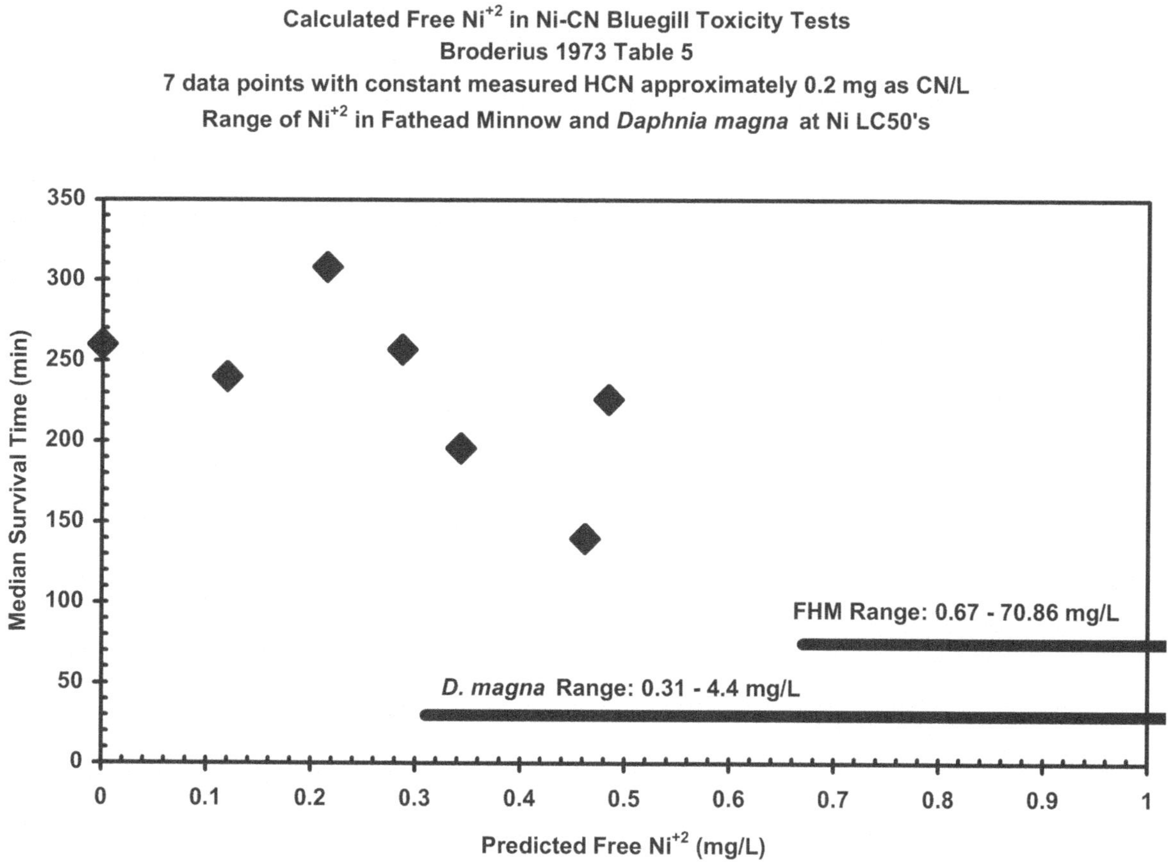

Figure 2-3. Median Survival Time of Bluegills In the Presence of Nickel-Cyanide Mixtures is Relatedto Free Nickel Ion Concentrations. For comparison, the range of free nickel ion concentrations for two other aquatic organisms, *D. magna* and fathead minnow, calculated by the Biotic Ligand Model are shown.

For the Ag-CN tests, threespine sticklebacks were exposed to either free cyanide, or a mixture of Ag-CN species and free cyanide. Chemical speciation calculations indicated a considerable amount of free silver ion in these test solutions. The additional toxicity observed with the Ag-CN and free cyanide mixture as measured by the difference in free cyanide between HCN only and Ag-CN/HCN solutions (i.e., the distance between the trendlines in Figure 2-2) is well correlated with the amount of silver bound to the biotic ligand as calculated by the Biotic Ligand Model (Figure 2-4). The additional toxicity, therefore, can be explained by silver ion toxicity and not necessarily by toxicity due to Ag-CN complexes.

For both the Ni-CN and Ag-CN exposures, therefore, bioavailability modeling suggests that sufficient metal ions were present in the exposures to explain the added toxicity. These experimental results are consistent with the prevailing assumption throughout this document that free cyanide is the most toxic form of cyanide, and that metal-cyanide complexes are relatively non-toxic in comparison.

Figure 2-4. Additional Toxicity Observed In Ag-CN Mixtures to Threespine Sticklebacks is Related to the Amount of Ag Bound to the Gill Predicted by the Biotic Ligand Model.

2.2 Cyanide Analytical Testing Methods

Various analytical methods are available that measure total cyanide, free cyanide, metal-cyanide species, or free cyanide plus easily dissociated complexes (Ghosh et al., 2006). The conventional total cyanide test involves acidifying a sample to pH < 2 and heating it to dissociate metal cyanide complexes, followed by distillation to liberate the HCN which is then captured in a sodium hydroxide (NaOH) trap (American Public Health Association, 1998). This method measures cyanide present as HCN, CN^-, and as metal-cyanide complexes. It does not measure any organocyanide that may be present, and it does not distinguish among the inorganic forms of cyanide. Weak acid dissociable (WAD) cyanide is measured by a similar distillation method, but only weak acid is added to the sample and the distillation pH is usually in the range of 4.5-6 (American Public Health Association, 1998; Zheng et al., 2003). Free cyanide can be measured by weakly acidifying a sample and measuring HCN that volatilizes at room temperature. The HCN is captured by contacting the enclosed headspace above the sample with an NaOH trap (ASTM, 1998a). The above are published, standard methods for determining cyanide concentrations.

In the majority of the studies used in the AWQC document for cyanide (U.S. EPA, 1985b), free cyanide was determined using the pyridine-pyrazalone colorimetric method from *Standard Methods* (American Public Health Association, 1971) with direct colorimetric development and no distillation. Normally, this method includes a key distillation step that not only reduces or eliminates interferences, but also "converts most cyanide complexes into the simple CN^- that is readily measured by titration or color test" (American Public Health Association, 1971). The method introduction specifies that distillation can be omitted if "it is known that the sample contains only simple cyanides of the alkalis and is completely free of all interferences" (American Public Health Association, 1971). The same document has a list of interferences that includes sulfides, heavy metal ions, fatty acids, thiocyanate, cyanate, glycine, urea, and oxidizing agents.

In the method description, it is clearly explained that the spectrophotometric (i.e., colorimetric) test is sensitive to variations in salt concentration. When working with saline solutions, the method states that it is "essential to isolate these salts so that a controlled salt concentration may be used for all aliquots of the distillate and the standards, as provided in the colorimetric test" (American Public Health Association, 1971). Thus, inaccuracies in the direct colorimetric procedure, without distillation, can occur due to variable ionic strengths among samples. If the free cyanide concentration of water containing other constituents is being tested, then that water should be used for the calibration curve, rather than deionized water. Thus, inclusion of the distillation is recommended to eliminate variations due to differences in water composition, and the accuracy of the direct colorimetric procedure without distillation is uncertain.

Another source of error for the direct colorimetric procedure is pH variation, which will affect the ion distribution. Any change in the water composition that will interfere with the spectrophotometric procedure should be accounted for when constructing the calibration curve. Additionally, metal-cyanide species can also interfere with the spectrophotometric method, in that they may dissociate during sampling and sample storage. Thus, for samples dominated by metal-cyanide complexes, it will be difficult to quantify the free cyanide content of the sample with a reasonable degree of accuracy, without first separating the complexes from the solution to be analyzed for free cyanide.

2.3 Evaluation of Cyanide Analyses Used in Toxicity Tests

Given that free cyanide is the most toxicologically relevant form of cyanide, toxicity studies should preferably use a reliable analytical method for free cyanide to quantify exposure concentrations in test solutions. However, as various analytical methods have been used in toxicity studies to measure cyanide, the link between exposure concentration and response in the studies may not be considered accurate or reliable. Further, the accuracy of cyanide analyses varies by method, making comparisons among concentration-response data difficult. Therefore, a classification scheme was developed to reflect the type of cyanide exposure measurements performed in the studies upon which the 1984 AWQC for cyanide are based, and whether free cyanide was measured or could be estimated using thermodynamic calculations. This classification scheme was developed not necessarily to "rank" scientific reliability, but rather to classify each study with respect to how (or whether) cyanide exposure chemistry was analytically measured. Ultimately, this analytical method evaluation for each toxicity study does not result in the removal of any toxicity data from the criteria calculation, but the evaluation is used to indicate the level of confidence in the toxicity studies most critical to the criteria values.

The cyanide analysis classification scheme used in evaluating existing aquatic toxicity test data (Tables 1a, 1b, 2a, and 2b) was as follows:

1. **Measured Free Cyanide:** A classification of 1 was assigned to studies for which a reliable (i.e., of known accuracy) method was used to measure free cyanide, such as Free Cyanide by Microdiffusion (ASTM 4282-95), or using a cyanide ion selective electrode (*Standard Methods* 4500 CN^- F).

2. **Measured Free Cyanide, but Accuracy of the Measurement is Unknown:** A classification of 2 was assigned to studies for which an approximate method was used to measure free cyanide, such as the *Standard Methods* 4500 CN^- E, but without the distillation. This includes analyses such as Free Cyanide by Epstein colorimetric method (American Public Health Association, 1971), the pyridine-pyrazolone colorimetric method, and any colorimetric method that does not include a distillation. While in the past it has been assumed that this was a reliable measurement of free cyanide, its accuracy compared to more recent techniques (e.g., microdiffusion) is not well understood (Ghosh et al., 2006).

3. **Measured Weak Acid Dissociable (WAD) Cyanide:** A classification of 3 was assigned to studies in which weak acid dissociable (WAD) cyanide was measured. WAD measurement includes free cyanide as well as weak metal-cyanide complexes such as zinc, nickel, copper and cadmium cyanide complexes. Methods for measuring WAD include *Standard Methods* 4500 CN^- I.

4. **Measured Total Cyanide and Sufficient Water Composition Information Provided to Estimate Free Cyanide Using Chemical Equilibrium Modeling:** A classification of 4 was assigned to studies in which total cyanide was measured, and detailed water composition information was included. Total cyanide testing methods include *Standard Methods* 4500 CN- E, and EPA Method 335.2. Sufficient water quality information includes major ion concentrations, metal concentrations, pH, and temperature. Specific ions needed for this purpose are compiled in Table 2-1 below. This information is sufficient to enter into a chemical equilibrium model and estimate the free cyanide concentration.

Table 2-1. Total Element/Ion Data Needed for Equilibrium Modeling of Cyanide Speciation Based on Total Cyanide Measurements.

Al	Ba	Be	B	Cd	Cr	Fe
Ni	Cu	Pb	Zn	Sr	H_3BO_3	Mg
Mn	Mo	Ca	Na	P	PO_4	K
Si	HCO_3	Cl	F	Br	SO_4	NH_3

5. **Measured Total Cyanide and Insufficient Water Composition Information Provided to Estimate Free Cyanide Using Chemical Equilibrium Modeling:** A classification of 5 was assigned to studies in which total cyanide was measured, but detailed water composition was not reported. Total cyanide testing methods include *Standard Methods* 4500 CN- E, and EPA Method 335.2. In this case, free cyanide concentrations cannot be estimated with the aid of chemical equilibrium modeling.

6. **No Analytical Verification of Cyanide Concentration:** A classification of 6 was assigned to studies in which no form of cyanide was measured, or no analytical methods were reported in the study.

Balancing scientific reliability and practicality, studies with Classifications 1-4 are probably most appropriate for use in deriving AWQC. This is because these studies either provide empirical measures of free cyanide, or thermodynamic calculations could be used to estimate free cyanide with reasonable confidence. Studies with Classifications 5 or 6 are much less reliable since it is not possible to estimate free cyanide from the data provided or because cyanide was not measured at all. However, when only toxicity data with Classifications 5 or 6 were available for a test aquatic species, they were included in the criteria calculations.

2.4 Expression of Cyanide Concentrations

Consistent with the 1984 AWQC document for cyanide, all cyanide concentrations reported in this report are expressed as CN ion. If toxicity data reported in the original studies were expressed in terms of free cyanide as CN, then no adjustment was necessary. However, if the free cyanide concentration was expressed as HCN, KCN, etc., then the concentration was adjusted based on the relative molecular weights of the compound and CN. Finally, when cyanide concentrations were reported in the original literature as HCN, rather than in terms of free cyanide, the data were converted from molecular HCN to free cyanide ($HCN + CN^-$) as CN using the following equation:

$$\mu\text{g free cyanide as CN/L} = (\mu\text{g HCN/L})\,(1 + 10^{\text{pH-pK}_{\text{HCN}}}) \times \frac{\text{mol. wt. CN}}{\text{mol. wt. HCN}} \tag{2-1}$$

where:

$$\text{pK}_{\text{HCN}} = 1.3340 + \frac{2347.2}{\text{T} + 273.16}$$

and T = degrees celsius

2.5 Methods for Updating Cyanide AWQC

Because recalculations of cyanide AWQC build from principles set forth in the 1985 U.S. EPA Guidelines (U.S. EPA, 1985b), it is useful to have some understanding of how the Guidelines are ordinarily applied: 1) Acute toxicity test data must be available for aquatic species from a minimum of eight diverse taxonomic groups. The diversity of tested species is intended to ensure protection of various components of an aquatic ecosystem. 2) The final acute value

(FAV) is derived by extrapolation or interpolation to a hypothetical genus more sensitive than 95% of all tested genera. The FAV, which represents an LC50 or EC50, is divided by two in order to obtain an acute criterion protective of nearly all individuals in such a genus. 3) Chronic toxicity test data (longer-term survival, growth, or reproduction) must be available for at least three taxa. Most often, as is the case for cyanide, the chronic criterion is set by determining an appropriate acute-chronic ratio (the ratio of acutely toxic concentrations to chronically toxic concentrations) and applying that ratio to the FAV. 4) When necessary, the acute and/or chronic criterion may be lowered to protect recreationally or commercially important species. As presented later, for cyanide the existing acute and chronic freshwater criteria were set lower to protect rainbow trout. 5) When evaluating time-variable ambient water concentrations generally a 1-hour average concentration is considered to be appropriate for comparison with the acute criterion, and a 4-day average concentration with the chronic criterion. 6) The allowable frequency for exceeding a criterion is set at once every three years on average.

CHAPTER 3.0

TOXICITY TO FRESHWATER ANIMALS

The toxicity of cyanide to freshwater animals has most recently been reviewed in Gensemer et al. (2006). This review included toxicity studies referenced in the AWQC for cyanide (U.S. EPA, 1985b) and updated the cyanide toxicity database with studies published since 1984. All of the studies discussed in Section 3.0 below would be considered acceptable for use in AWQC derivation according to U.S. EPA Guidelines (U.S. EPA, 1985b).

3.1 Acute Toxicity

The 1984 AWQC (U.S. EPA, 1985a) included acute cyanide toxicity data for seven species and six genera of invertebrates and ten species and nine genera of fish. Invertebrate GMAVs ranged from 123.6 µg/L for *Daphnia* spp. to 2,490 µg/L for the midge *Tanytarsus dissimilis*. For fish, GMAVs ranged from 63.45 µg/L for *Salmo* spp. to 318 µg/L for the goldfish *Carassius auratus*. The updated acute toxicity data for freshwater organisms are provided in Table 1a. Toxicity data published since 1984 (Gensemer et al., 2006) increased the number of invertebrate species from 7 to 17 and the number of invertebrate genera from 6 and 14. The number of fish species increased to 11 and the number of genera increased from 9-11, due to the reclassification of rainbow trout from the genus *Salmo* to *Oncorhynchus* and because the additional fish species was in a genera not previously represented.

Most of the toxicity studies with invertebrates were given Classification 6 (i.e., cyanide concentration not measured). However, the studies used to derive the GMAVs for the four most sensitive genera were given Classification 1 or 2, which indicate that toxicity results are based on empirical measures of free cyanide. The ranked GMAVs are provided in Table 3a. The cyanide toxicity data for the four most sensitive genera are summarized below:

<u>Genus</u>: *Oncorhynchus*

The *Oncorhynchus* GMAV of 46.53 µg CN/L is based exclusively on toxicity data for rainbow trout (*O. mykiss*). Of the 14 LC50s meeting requirements for inclusion in Table 1, seven were given Classification 2 and seven were given Classification 6. As discussed above, preference was given to studies with Classifications 1-4, so the GMAV was calculated based on those studies with Classification 2. The LC50s comprising the GMAV range from 27 to 57 µg CN/L. Kovacs (1979) observed that the LC50 decreased with decreasing temperature, with LC50 values of 27, 40, and 65 µg CN/L at temperatures of 6, 12, and 18°C, respectively (to be conservative, all LC50s were included in calculation of the rainbow trout SMAV). This revised SMAV for rainbow trout differs slightly from the SMAV of 44.73 µg CN/L in the 1984 AWQC due to publication of a new study (McGeachy and Leduc, 1988).

Genus: *Salvelinus*

The *Salvelinus* GMAV of 85.80 µg CN/L is based exclusively on toxicity data for brook trout (*Salvelinus fontinalis*) and is unchanged from the 1984 AWQC. The *S. fontinalis* GMAV is based on 20 LC50s for sensitive life stages (swim-up fry, juveniles), which ranged from 52 to 156 µg CN/L (Cardwell et al., 1976; Smith et al., 1978) and all were given a classification number of 2, which is considered acceptable for use in deriving free cyanide-based criteria.

Genus: *Salmo*

The *Salmo* GMAV of 90.00 µg/L is based on a single Atlantic salmon (*S. salar*) LC50 of 90.00 µg CN/L (Tryland and Grande 1983) and is unchanged from the 1984 AWQC. This study was given Classification 1, which is considered acceptable for use in deriving free cyanide-based criteria.

Genus: *Perca*

The *Perca* GMAV of 92.64 µg CN/L is based exclusively on toxicity data for yellow perch (*Perca flavescens*) and is unchanged from the 1984 AWQC. The *P. flavescens* GMAV is based on nine LC50s, which ranged from 74.7 to 107 µg CN/L (Smith et al., 1978) and all were given Classification 2, which is considered acceptable for use in deriving free cyanide-based criteria.

Freshwater Acute Criterion Calculation

Based on the four lowest GMAVs and a total number of GMAVs of 28, the estimated 5th percentile (FAV) following the U.S. EPA Guidelines (U.S. EPA, 1985b) was estimated to be 58.95 µg CN/L (Table 3c). This value is greater than the GMAV of 46.53 µg CN/L for rainbow trout, which is a recreationally and economically important species. Accordingly, the FAV should be lowered to 46.53 µg CN/L. Division of this value by two results in a CMC of 23.27 µg CN/L (Table 3c, Figure 3-1a). When a criterion is driven by the sensitivity of coldwater species, a site-specific option for developing a criterion applicable to warmwaters is to remove GMAVs for coldwater species and recalculate the FAV (e.g., the 1999 Update to the ammonia AWQC; U.S. EPA, 1999). For cyanide, salmonids constitute the three most sensitive genera. Removing the salmonid GMAVs results in the following as the four lowest GMAVs: 92.64 µg CN/L (*Perca flavescens*); 98.80 µg CN/L (*Gasterosteus aculeatus*); 99.28 µg CN/L (*Lepomis macrochirus*); and 102.0 µg CN/L (*Micropterus salmoides*). With a reduced N of 25, the FAV for warmwater species would be 94.50 µg CN/L and the resulting warmwater CMC would be 47.25 µg CN/L (Table 3c, Figure 3-1b). However, eliminating coldwater species from the criteria calculation should be done with caution when threatened or endangered (T&E) species occur in the water body (see Section 8).

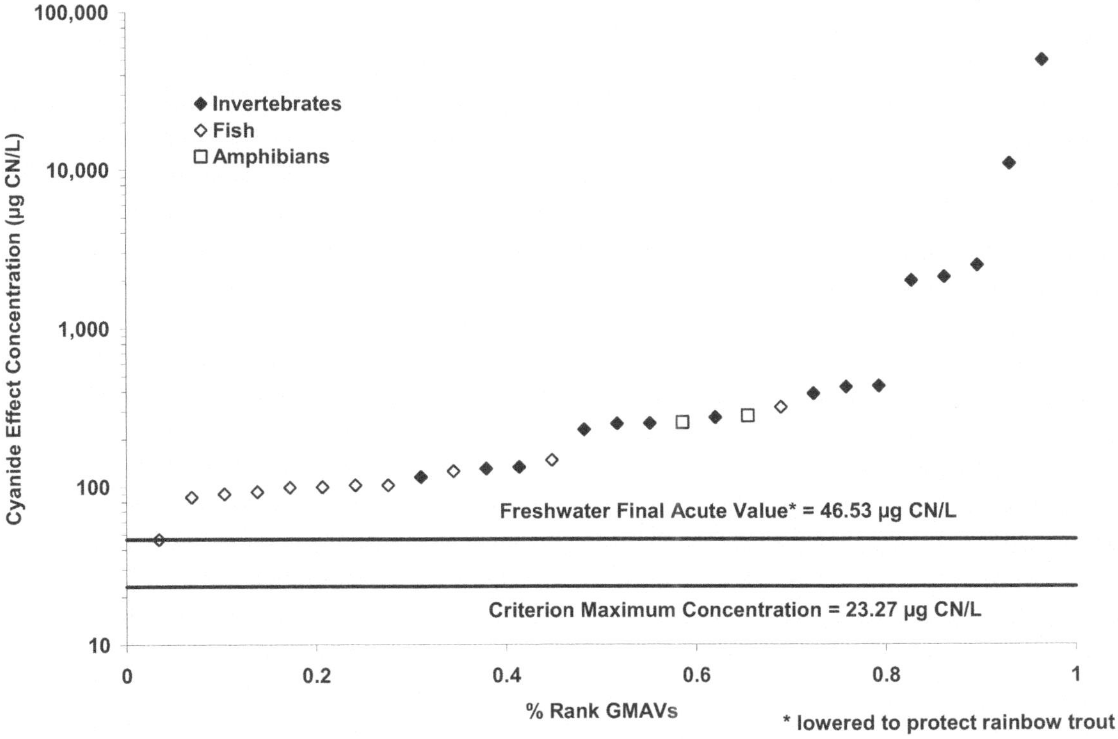

Figure 3-1a. Ranked Summary of Cyanide GMAVs (freshwater, all species).

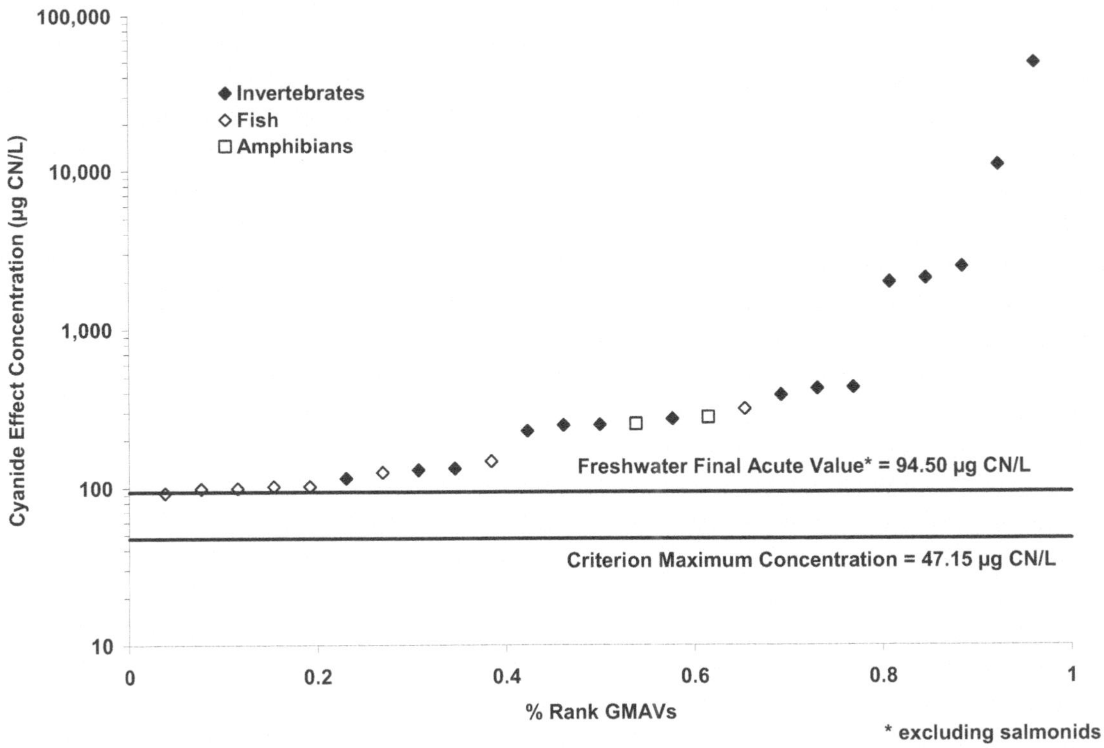

Figure 3-1b. Ranked Summary of Cyanide GMAVs (freshwater, excluding salmonids).

3.2 Chronic Toxicity

The 1984 AWQC (U.S. EPA, 1985a) included chronic cyanide toxicity data for two species of invertebrates and three species of fish. Chronic values, defined in the 1984 AWQC as the geometric mean of the no observed effect concentration (NOEC) and lowest observed effect concentration (LOEC), ranged from 7.849 µg/L for the brook trout *S. fontinalis* (Koenst, et al. 1977) to 34.06 µg/L for the isopod *Asellus communis* (Oseid and Smith 1979; see also Table 2a). No additional chronic studies for freshwater species that meet the Guidelines were identified. All chronic studies were given Classification 2, which is considered acceptable for use in deriving free cyanide-based criteria.

In the recalculated cyanide AWQC presented in this report, chronic EC20s were calculated using regression analysis so that the chronic values are associated with a uniform level of effect. This approach is consistent with U.S. EPA Guidelines (U.S. EPA, 1985b), which states that the chronic value can be calculated using hypothesis testing (i.e., NOEC, LOEC) or regression analysis. This is also consistent with more recent AWQC updates such as the 1999 update for ammonia (U.S. EPA, 1999) and the draft updated AWQC for copper (U.S. EPA, 2003a) which selected the EC20 as an appropriate compromise between aquatic life protection and statistical precision. A hierarchical approach was used for estimating the EC20 for each study:

(1) The first preference was to use the U.S. EPA's Toxicity Relationship Analysis Program (TRAP) to calculate EC20 values. This program analyzes the decline in a biological variable from its control value to zero as the chemical concentration increases. The user can select one of six statistical models to evaluate the data. These include three probability distribution models (normal, uniform, triangular) and three nonlinear regression models (logistic, piecewise-linear, piecewise-tailed). The TRAP help manual cautions that the parameter estimates and confidence limits should be examined relative to the data to determine whether they are reasonable. If an EC20 estimated by TRAP was not concordant with the raw toxicity data, alternative methods below were used for estimating the EC20.

(2) If an EC20 could not be estimated using the TRAP software, other standard distributions (e.g., log-probit) were fit to the concentration-response data using commercially available software to estimate an EC20 value. As for EC20s estimated using the TRAP software, EC20s estimated from other distribution fitting techniques were visually compared to the raw toxicity data and discarded if they were not concordant with the raw toxicity data.

(3) If an EC20 could still not be reasonably estimated using other distribution fitting software, the EC20 was estimated using linear interpolation.

The following summarizes the chronic toxicity data for freshwater species and presents the regression analyses for determining EC20 values.

Asellus communis (isopod)

The 1984 AWQC (U.S. EPA, 1985a) cited a cyanide NOEC and LOEC of 29 and 40 µg CN/L, respectively, for the isopod *Asellus communis* (Oseid and Smith, 1979). Although the authors did not test for statistical significance between cyanide treatment and control responses, mean weights of organisms exposed to cyanide concentrations of 40 µg CN/L and higher were lower than the mean weight of the controls by more than two times the standard error of the mean (weight was based on the total weight of free individuals plus eggs and young). The mean weights of organisms exposed to cyanide concentrations of 29 µg CN/L and lower were greater than the mean control weight. For determination of the EC20, the weight data for each cyanide treatment were converted to the percent reduction in weight relative to the control. The TRAP model was unable to estimate an EC20 and due to the concentration-response data being somewhat anomalous (i.e., the reduction in growth does not become incrementally greater with increasing cyanide concentration), other distribution types also did not provide a reasonable estimate of the EC20 value. For example, a log-probit model was fit to the data (Figure 3-2), which provided an estimate of the EC20 with a very large confidence interval (61-3,089 µg CN/L). Further, the log-probit model estimates an EC20 of 26.5 µg CN/L that is less than the highest cyanide concentration (29 µg CN/L) that resulted in a zero percent reduction in growth relative to the control. Consequently, a simpler, but perhaps more accurate, approach for estimating the EC20 is to linearly interpolate between the NOEC (0% reduction in growth relative to the control) and the LOEC (83% reduction in growth relative to the control). This results in an estimated EC20 of 32 µg CN/L.

Gammarus pseudolimnaeus (amphipod)

The 1984 AWQC (1985a) cited a cyanide NOEC and LOEC of 16 and 21 µg CN/L, respectively, for the amphipod *Gammarus pseudolimnaeus* (Oseid and Smith, 1979). The highly variable responses in the two control tanks somewhat confound interpretation of the results, but the mean number of eggs or young per gravid female was reduced to zero at cyanide concentrations of 21 µg CN/L and higher. At cyanide concentrations of 16 µg CN/L and lower, the mean number of eggs or young per gravid female is comparable to the mean number in the control tanks. It should be noted that the mean number of eggs and young per gravid female in the two control tanks were 3 and 27 – the basis for this variability is unclear. Nevertheless, these highly different control values were averaged for the purpose of deriving an EC20. As for *A. communis*, an EC20 could not be estimated using the TRAP software. A log-probit model was fit to the data, resulting in an estimated EC20 of 14.31 µg/L (Figure 3-3). Thus, the EC20 is lower than the chronic value of 18.33 µg/L cited in the 1984 AWQC.

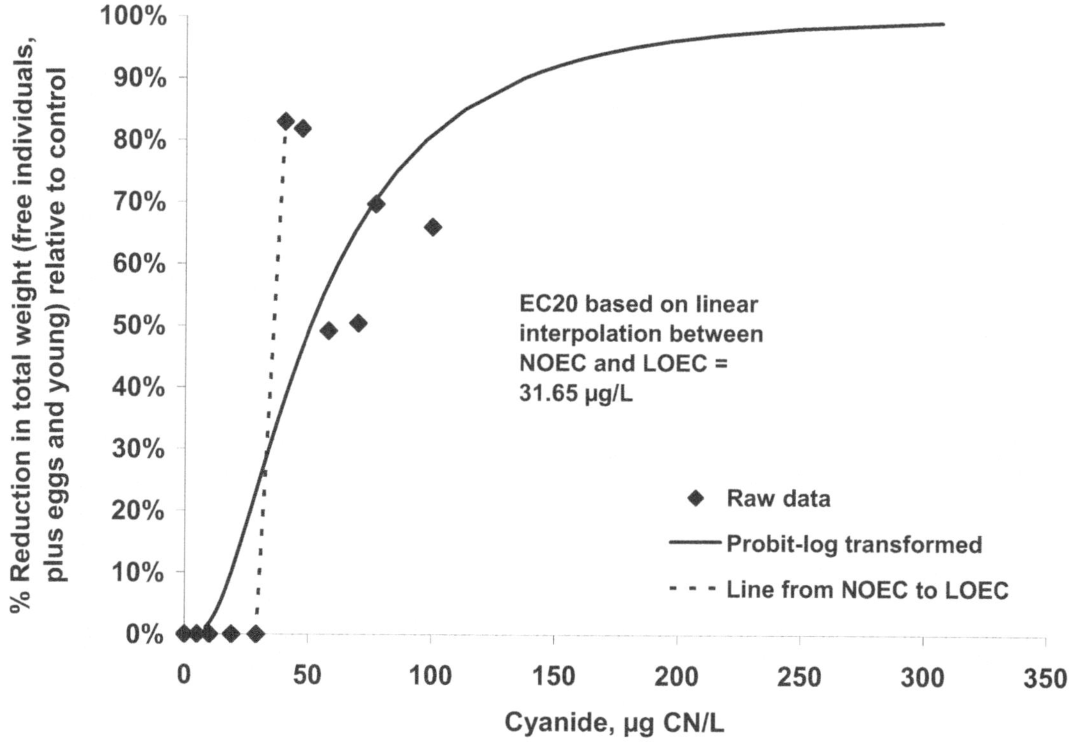

Figure 3-2. Concentration-response Data for *Asellus communis* Exposed to Cyanide. Data from Oseid and Smith (1979).

Salvelinus fontinalis (brook trout)

The 1984 AWQC (U.S. EPA, 1985a) cited a cyanide NOEC and LOEC of 5.6 and 11.0 µg CN/L for the brook trout *Salvelinus fontinalis* (Koenst, et al., 1977). These NOEC and LOEC values were based on the mean number of eggs spawned per female. To derive an EC20, the data for mean number of eggs spawned per female were combined with percent egg viability data to calculate the mean number of viable eggs spawned per female. These data for the cyanide treatments were then converted and expressed as the percent reduction relative to the control. As for *A. communis* and *G. pseudolimnaeus*, the TRAP software was unable to estimate an EC20. A log-probit model was fit to the data (Figure 3-4), but because the concentration-response is anomalous at a cyanide concentration of 43 µg CN/L and due to the somewhat unusual shape of the curve (steeper at the lower and upper tails), the log-probit model appears to greatly overestimate toxicity in the lower tail. Consequently, the EC20 was linearly interpolated from the NOEC (2% reduction in viable eggs spawned per female relative to the control) and the LOEC (52% reduction in viable eggs spawned per female relative to the control). This results in an estimated EC20 of 7.705 µg CN/L, which is incidentally very similar to the EC20 estimated using the log-probit model (the line between the NOEC and LOEC happens to cross the log-probit model near the EC20).

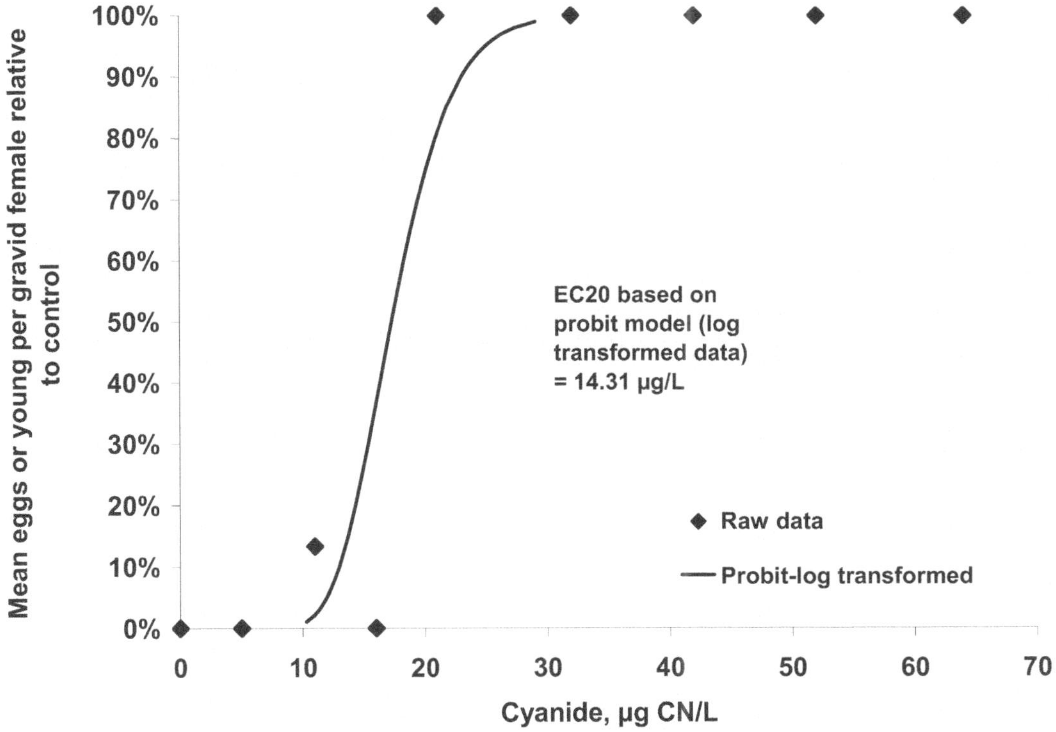

Figure 3-3. Concentration-response Data for *Gammarus pseudolimnaeus* exposed to Cyanide. Data from Oseid and Smith (1979).

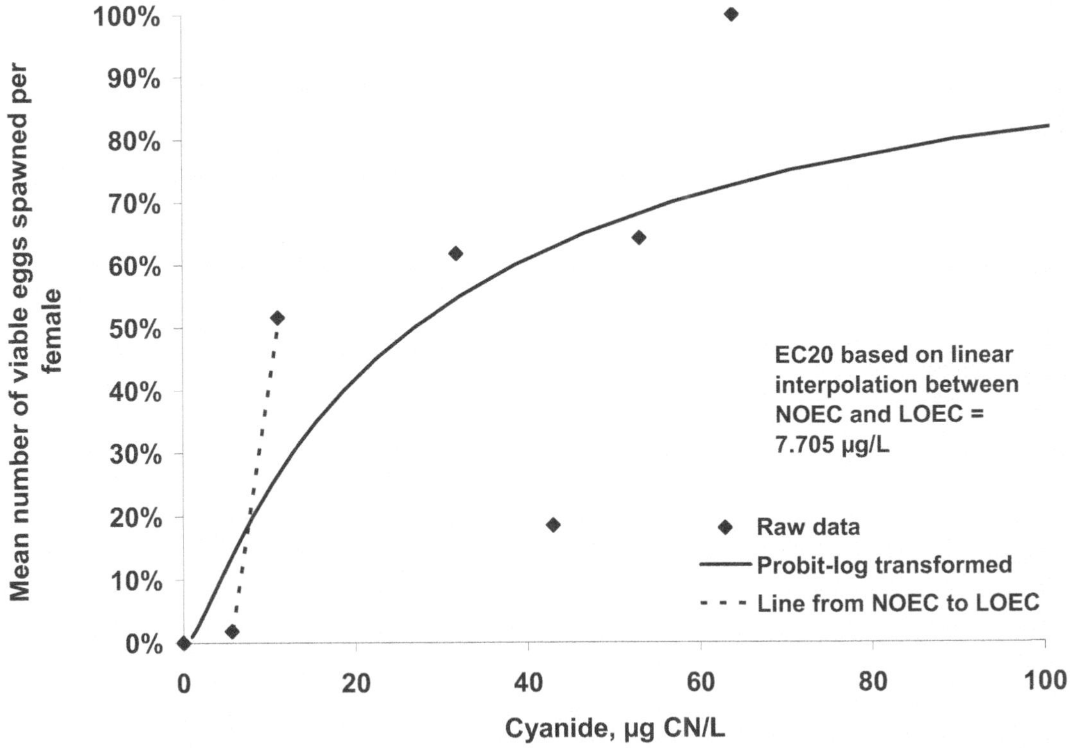

Figure 3-4. Concentration-response Data for *Salvelinus fontinalis* exposed to Cyanide. Data from Koenst et al. (1977).

Pimephales promelas (fathead minnow)

The 1984 AWQC (U.S. EPA 1985a) cited a cyanide NOEC and LOEC of 13.3 and 20.2 µg CN/L for the fathead minnow *Pimephales promelas* (Lind, et al., 1977). These NOEC and LOEC vales were based on the mean number of eggs per female. To derive an EC20, the data for the cyanide treatments were converted and expressed as the percent reduction in eggs per female relative to the control. The TRAP software estimated an EC20 of 14.56 µg CN/L; however, this is greater than the cyanide concentration that resulted in a 47% in the mean number of eggs per female (Figure 3-5). Alternatively, a log-probit model was fit to the data, resulting in an estimated EC20 of 6.038 µg CN/L that is more consistent with the raw data (Figure 3-5). This EC20 is lower than the chronic value of 16.39 µg CN/L cited in the 1984 AWQC.

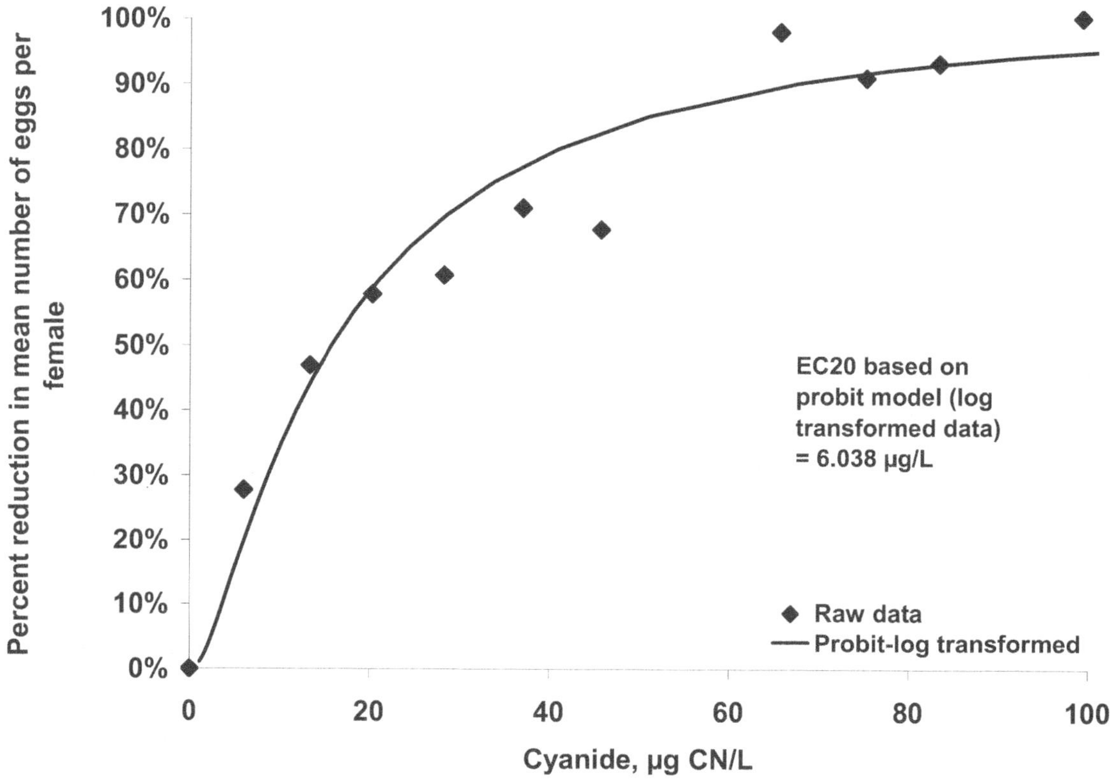

Figure 3-5. Concentration-response Data for *Pimephales promelas* Exposed to Cyanide. Data from Lind et al. (1977).

Lepomis macrochirus (bluegill)

The 1984 AWQC (U.S. EPA, 1985a) cited a cyanide NOEC and LOEC of 9.3 and 19.8 µg CN/L, respectively, for the bluegill *Lepomis macrochirus* (Kimball et al., 1978). These NOEC and LOEC values were based on survival of bluegills from fertilized eggs to 57-d old juveniles. To derive an EC20, the data for the cyanide treatments were converted and expressed as the percent reduction in survival relative to the control. As shown, in Figure 3-6, the 52.1 µg CN/L treatment was treated as an outlier based on visual inspection of the data (and the TRAP software was unable to estimate an EC20 with this treatment included in the analysis). The TRAP software estimated an EC20 of 8.941 µg CN/L. For comparison, a log-probit model was fit to the data, resulting in an EC20 estimate of 5.595 µg CN/L (Figure 3-6). Because these values are reasonably similar, and both provide reasonable estimates of the EC20 when compared to the raw toxicity data, the EC20 of 8.941 µg CN/L estimated by the TRAP software was selected. This EC20 is lower than the chronic value of 13.57 µg CN/L cited in the 1984 AWQC.

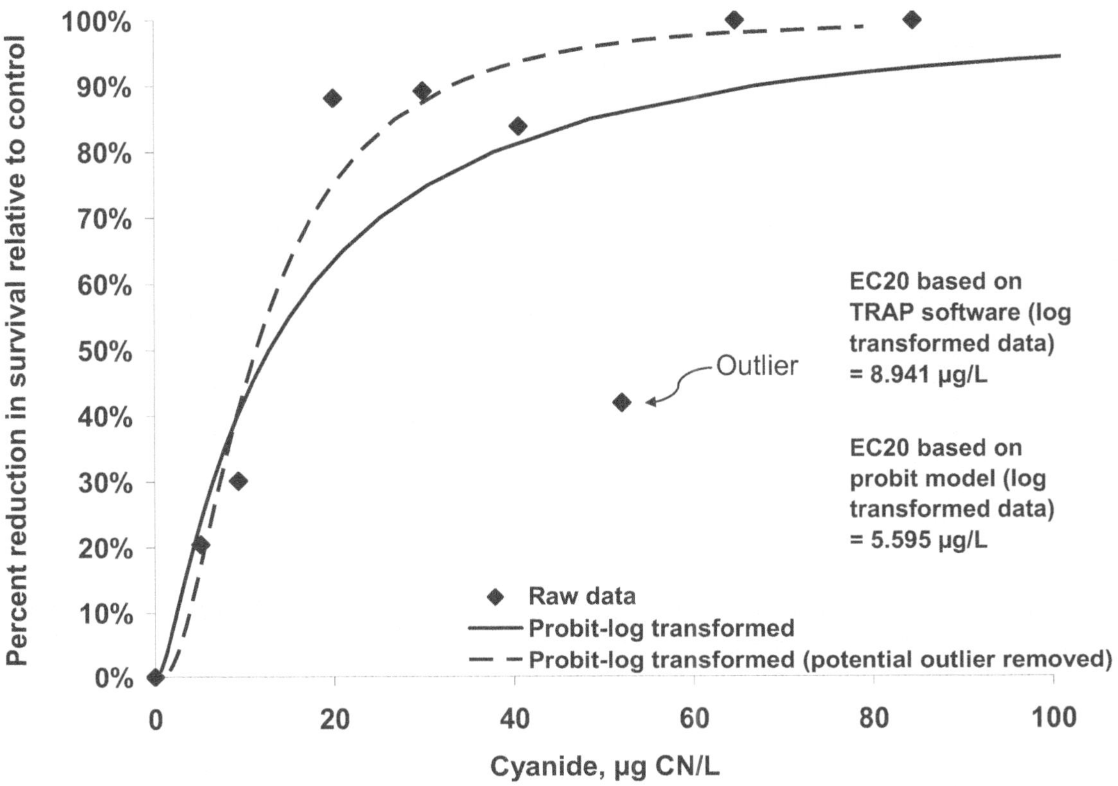

Figure 3-6. Concentration-response Data for *Lepomis macrochirus* Exposed to Cyanide. Data from Kimball et al. (1978).

The chronic values reported in the 1984 AWQC, which are based on the geometric mean of the NOEC and LOEC, are compared to the chronic values based on EC20 values in Figure 3-7. The overall trend of EC20 values resulting in lower chronic values results in substantial changes to the acute-chronic ratios (ACRs) relative to the 1984 AWQC for cyanide. In the 1984 AWQC, the geometric mean ACR for freshwater species was calculated to be 8.568. This ACR was based on ACRs of 9.111 for the amphipod *G. pseudolimnaeus*, 10.59 for brook trout (*S. fontinalis*), 7.633 for the fathead minnow (*P. promelas*), and 7.316 for bluegill (*L. macrochirus*). The ACR of 68.21 for the isopod *A. communis* was not included in the 1984 AWQC because it was much greater than those for more sensitive species; the researchers also recommend not using this ACR for the same reasons. In the Guidelines (U.S. EPA, 1985b), it is stated that "if no major trend is apparent and the acute-chronic ratios for a number of species are within a factor of ten, the Final Acute-Chronic Ratio should be calculated as the geometric mean of all the species mean acute-chronic ratios available for both freshwater and saltwater species." Accordingly, using the chronic EC20s to derive the ACRs, and including ACRs for saltwater species (discussed in Section 4 below), results in a Final ACR (FACR) of 9.659 (Table 2c).

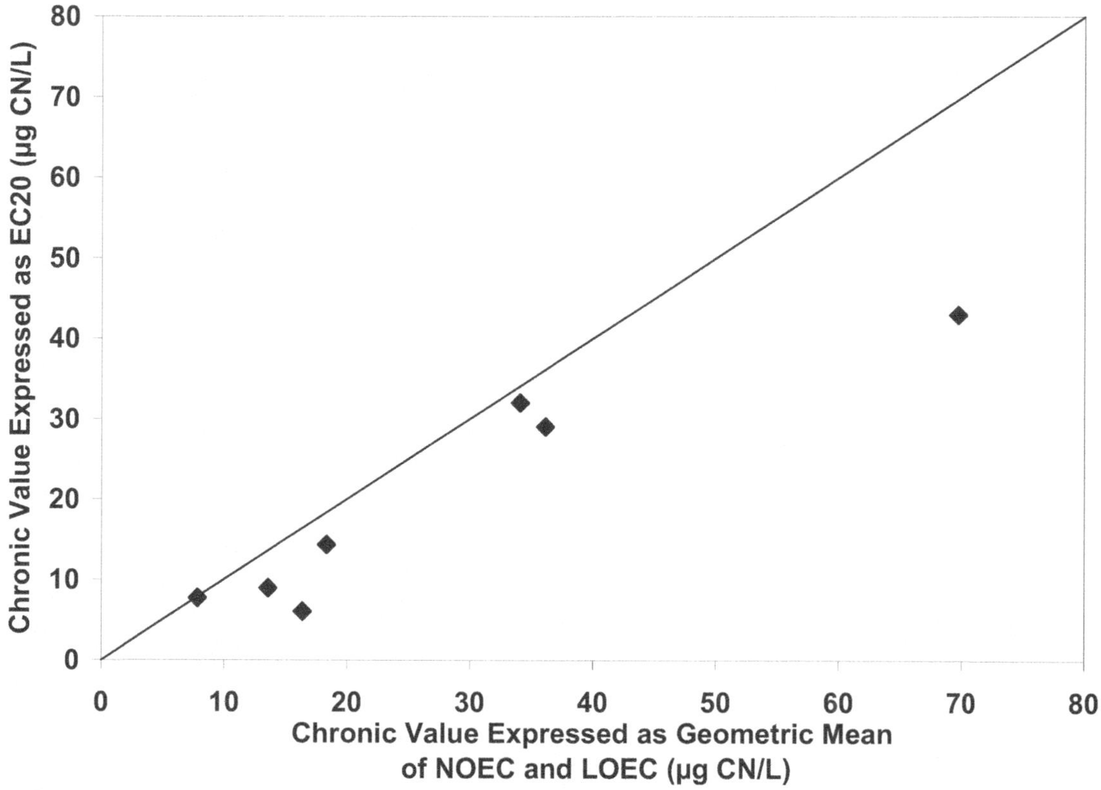

Figure 3-7. Comparison of Chronic Values Expressed as the Geometric Mean of the NOEC And LOEC Vs. Chronic Values Expressed as EC20s. Solid Line Represents Perfect Agreement.

<u>Freshwater Chronic Criteria Calculation</u>

Because chronic cyanide toxicity data are not available for the minimum number of families, the FCV was estimated from the FAV using an acute-chronic ratio (ACR). As discussed above, the geometric mean FACR for cyanide is 9.659 when chronic EC20 values are used. Division of the rainbow trout-based FAV of 46.53 µg/L by the FACR of 9.659 results in a FCV of 4.817 µg CN/L (Table 3c, Figure 3-8). If the FAV based on warmwater species is considered, the resulting FCV is 9.784 µg CN/L (Table 3c).

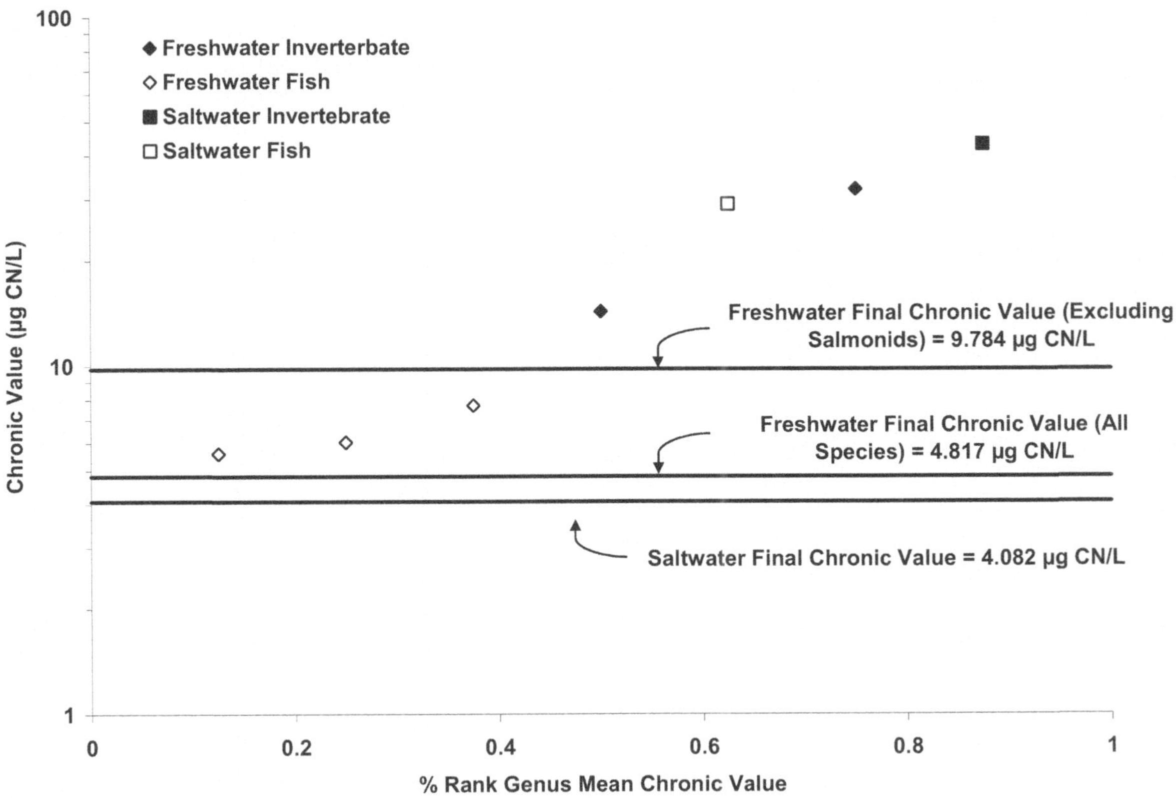

Figure 3-8. Chronic Toxicity of Cyanide to Aquatic Animals.

CHAPTER 4.0

TOXICITY TO SALTWATER ANIMALS

The toxicity of cyanide to saltwater animals has most recently been reviewed in Gensemer et al. (2006). This review included toxicity studies referenced in the AWQC for cyanide (U.S. EPA, 1985b) and updated the cyanide toxicity database with studies published since 1984. All of the studies discussed in Section 4 below would be considered acceptable for use in AWQC derivation according to U.S. EPA Guidelines (U.S. EPA, 1985b).

4.1 Acute Toxicity

The 1984 AWQC (U.S. EPA, 1985b) include cyanide toxicity data for six species and five genera of invertebrates, and three species and genera of fish (Table 1b). For invertebrates, GMAVs ranged from 17 μg CN/L based on the copepod *Acartia clausi* to >10,000 μg/L based on the common Atlantic slippershell *Crepidula fornicata*. Fish GMAVs ranged from 59 μg/L based on the Atlantic silverside *Menidia menidia* to 372 μg/L based on the winter flounder *Pseudopleuronectes americanus*. Toxicity data published since 1984 increased the number of invertebrate species and genera to 11 and 6, respectively, but no additional cyanide toxicity data were identified for any additional fish species (Gensemer et al., 2006). It is important to note that the cyanide toxicity data for the copepod *Acartia clausi* were reanalyzed by the U.S. EPA and the LC50 was revised from 30 μg CN/L (in the 1984 AWQC) to 17 μg CN/L (Lussier and Cardin, 1985; Hansen and Burgess, 1987). An independent analysis of the *A. clausi* toxicity data using a Trimmed Spearman Karber estimate confirmed an LC50 of 17 μg CN/L.

The ranked saltwater GMAVs are provided in Table 3b. The GMAV for the most sensitive genera, *Acartia*, is based on the *A. clausi* study mentioned above, which has a Classification 6 (cyanide was not measured), while the toxicity studies for the next three most sensitive genera are based on Classifications 2-4 (indicating reliable measures of free cyanide-based toxicity). As was done in the 1984 AWQC, the *A. clausi* study was retained in Table 1b even though cyanide concentrations were not measured because this was the only 96-hr acute study available for this genus. The cyanide toxicity data for the four most sensitive genera are summarized below:

<u>Genus</u>: *Acartia*

The *Acartia* GMAV of 17.00 μg CN/L is based on a single *A. clausi* LC50 of 17 μg CN/L, which, as mentioned above, was lowered from 30 μg CN/L (in the 1984 AWQC) based on a re-analysis of the data by the U.S. EPA (Lussier and Cardin, 1985; Hansen and Burgess, 1987). This study was given Classification 6 since cyanide concentrations were unmeasured.

The acute toxicity of cyanide to another *Acartia* species, *A. tonsa*, was recently evaluated using 48-hr static exposures (Parametrix, 2006). The exposure duration was reduced from 96 to

48-hr because preliminary tests showed that test organisms were unable to survive 96 hours without being fed. Although the U.S. EPA was able to conduct a 96-hour toxicity test with *A. clausi* without feeding the test organisms, keeping *Acartia* spp. unstressed and/or alive for 96 hours without feeding is not always feasible. In a recent example, Pedroso et al.(2007) found that whole body Na, K-ATPase activity was about 80% lower in *A. tonsa* which were fed relative to starved organisms, even at three different test salinities (5, 15, 30 ppt). Increased Na, K-ATPase activities in the unfed *A. tonsa* thus may reflect an attempt to increase ion uptake in the absence of food and, presumably, the increased energy requirements associated with trying to maintain proper ion balance. However, the researchers are not aware of any studies conclusively demonstrating that this additional stress could lead specifically to unacceptable control mortalities in unfed *A. tonsa* in acute tests lasting longer than 48-hr.

In the Parametrix cyanide studies, the 48-hour LC50s determined for *A. tonsa* were 195.0, 191.2, 217.1, and 234.7 µg CN/L, resulting in an *A. tonsa* SMAV of 208.8 µg CN/L. These toxicity studies were given Classification 1 because free cyanide was measured. However, because the exposure durations were not 96 hours, the data are not included in Table 1b and cannot be incorporated into the *Acartia* GMAV according to U.S. EPA guidance. Because it is not appropriate to directly compare 48-hr to 96-hr toxicity values, the 48-hour *A. clausi* LC50 was calculated to be 70 µg CN/L based on raw data from the original U.S. EPA studies (Walter Berry, U.S. EPA, Narragansett, RI, Personal Communication). This value is approximately 33% lower than the 48-hour *A. tonsa* SMAV of 208.8 µg CN/L, but a threefold difference in LC50 values between species within the same genus is not unreasonable (U.S. EPA, 1985b). Therefore, even if these *A. tonsa* data can not be used to recalculate the GMAV for *Acartia*, 48-hr SMAVs are close enough to suggest that the existing *A. clausi* 96-hr SMAV is a reasonable basis for the GMAV.

Genus: *Menidia*

The *Menidia* GMAV of 59 µg CN/L is based on a single *M. menidia* LC50 (Gardner and Berry, 1981) and is unchanged from the 1984 AWQC. This study was given Classification 2, as the endpoint is based on empirical measurements of free cyanide.

Genus: *Cancer*

In the 1984 AWQC for cyanide (U.S. EPA, 1985b), *Cancer* was the most acutely sensitive saltwater genus. The *Cancer* GMAV was based on a *C. irroratus* SMAV of 4.893 µg/L (Johns and Gentile, 1981), which was based on the geometric mean of two LC50s (4.2 and 5.7 µg CN/L). Brix et al. (2000) tested the sensitivity of West Coast *Cancer* species to develop a site-specific ambient water quality criterion for Puget Sound, Washington. West Coast *Cancer* species were found to be less sensitive to cyanide than *C. irroratus*, with SMAVs of 68.50 µg/L for *C. magister* (Dungeness crab), 130.7 µg CN/L for *C. oregonensis* (Oregon cancer crab), 143.7 µg CN/L for *C. gracilis* (slender crab), and 153.1 µg CN/L for *C. productus* (red crab). These SMAVs resulted in a GMAV of 118.5 µg CN/L for West Coast *Cancer* species. Recently, additional acute cyanide toxicity tests were conducted with the East Coast rock crab *C. irroratus*. Based on flow-through exposures and measurement of cyanide using microdiffusion, LC50s of 44.2, 70.4, and 70.9 µg CN/L were determined (Northwestern Aquatic Sciences, 2005c, 2005b, 2005a). Although the geometric mean of these three LC50s (60.4 µg CN/L) is over an order of magnitude greater than the *C. irroratus* GMAV in the 1984 AWQC document (4.893 µg/L), the basis for this discrepancy is unclear. Accordingly, all LC50s were combined to calculate an updated *C. irroratus* SMAV of 22.11 µg/L. Combining this with the West Coast *Cancer* SMAVs (Brix et al., 2000) results in a *Cancer* GMAV of 84.7 µg CN/L.

Genus: *Americamysis*

The *Americamysis* GMAV of 118.4 µg CN/L is based on an *A. bahia* LC50 of 133 µg CN/L (Lussier et al., 1985) and an *A. bigelowi* LC50 of 124 µg CN/L (Gentile, 1980), and is unchanged from the 1984 AWQC. The *A. bahia* and *A. bigelowi* studies were given classification numbers of 4 and 6, respectively, and so at least the *A. bahia* LC50 was based on measured cyanide concentrations.

Saltwater Acute Criteria Calculation

Based on the four lowest GMAVs and a total number of GMAVs of nine, the estimated 5th percentile following the Guidelines was estimated to be 11.00 µg CN/L (Figure 4-1). Division of this value by two results in a CMC of 5.500 µg CN/L (Table 3c). This recalculated CMC is approximately five times greater than the current CMC of 1.015 µg CN/L. The primary basis of this change is the increase of the *Cancer* GMAV from 4.893 µg CN/L to 84.69 µg CN/L. Because the number of GMAVs was less than 20, the 5th percentile (FAV) of 11.00 µg CN/L is an extrapolated value less than the lowest GMAV of 17.00 µg CN/L. Increasing the number of GMAVs generally increases the FAV to a value closer to the lowest GMAV, but the relative magnitudes of the four lowest GMAVs can also substantially influence the calculation of the FAV, sometimes resulting in greater extrapolation to the 5th percentile (in general, the more shallow the slope of the four lowest GMAVs, the greater the extrapolation to the 5th percentile).

4.2 Chronic Toxicity

The 1984 AWQC (U.S. EPA, 1985a) included chronic cyanide toxicity data for one saltwater invertebrate species (*Americamysis bahia*) and one saltwater fish species (*Cyprinodon variegates*). Chronic values, defined as the geometric mean of the NOEC and LOEC, were 69.71 µg CN/L for *A. bahia* and 36.12 µg CN/L for *C. variegates*. No additional chronic studies for saltwater species that meet the U.S. EPA Guidelines (U.S. EPA 1985b) were identified. Both of these chronic studies were assigned Classification 2, meaning the chronic values are based on empirical measurements of free cyanide. As was done for freshwater species, the chronic EC20 was calculated because it provides a more uniform measure of the chronic effect than the geometric mean of the NOEC and LOEC. Due to nuances with the chronic toxicity data for saltwater species, it was not possible to use the TRAP software or other distribution fitting techniques to reasonably estimate an EC20—see text below.

Americamysis bahia (mysid)

Lussier et al. (1985) reported no effect on chronic survival in *A. bahia* at the highest cyanide concentration tested (43 µg CN/L). In the 1984 AWQC (U.S. EPA, 1985a), the geometric mean of the chronic NOEC and acute LC50 of 113 µg CN/L was calculated to estimate the chronic value because no LOEC was available. Because the acute LC50 is likely not an appropriate estimate of the chronic LOEC, the chronic *A. bahia* data were re-evaluated to determine whether a chronic EC20 could be estimated. At the cyanide NOEC of >43 µg CN/L the number of young produced per female reproductive day was reduced by 23.8%. Even though this effect was not statistically different from controls, this is near a 20% effect level, and so the NOEC of 43 µg CN/L was assumed to provide a reasonable estimate of the cyanide EC20 for *A. bahia*.

Cyprinodon variegates (sheepshead minnow)

Based on the *C. variegates* study conducted by Schimmel et al. (1981), a NOEC and LOEC of 29 and 45 µg CN/L, respectively, were included in the 1984 AWQC (U.S. EPA, 1985a). Using Abbott's correction for control mortality, the percent mortality at the NOEC of 29 µg/L (the lowest concentration tested) was 23%. In the remaining cyanide concentrations tested, the percent mortality was very similar, ranging from 64-68% (Abbott's-normalized). Consequently, statistical models did not provide reasonable fits to the raw data and estimates of the EC20 appeared to be unreasonably low. Because the NOEC was associated with a 23% effect, it was assumed to provide a more reasonable estimate of the cyanide EC20 for *C. variegates*.

<u>Saltwater Chronic Criteria Calculation</u>

Because chronic cyanide toxicity data are not available for the minimum number of families, the FCV was estimated from the FAV using an ACR. In the 1984 AWQC for cyanide (U.S. EPA, 1985a), chronic criteria were driven by the acute sensitivity of *C. irroratus* larvae. However, the acute toxicity studies used to derive the available ACRs for saltwater organisms were based on the juvenile life stage of fish and a mysid. According to the 1984 AWQC, the acute toxicity value for rock crab larvae is probably a better indication of the chronic sensitivity of *C. irroratus* than would be obtained using the available ACRs. This is consistent with the U.S. EPA Guidelines (U.S. EPA, 1985b), which recommend using an ACR of 2 for acute embryo or larval toxicity studies with crabs and other saltwater invertebrates. However, following updates to the toxicity database for saltwater organisms, the most sensitive genus is now *Acartia*, rather than *Cancer*. Accordingly, it is possible that an ACR greater than 2 is now appropriate for determining the FCV. Following the U.S. EPA Guidelines (U.S. EPA, 1985b), the freshwater and saltwater ACRs can be combined into a geometric mean FACR if no major trend between ACRs and their corresponding acute values is apparent. For cyanide, this results in a geometric mean FACR of 9.659 (Table 2c). Application of this FACR to the saltwater FAV of 39.43 µg CN/L results in a FCV of 4.082 µg CN/L (Figure 3-8; Table 3c).

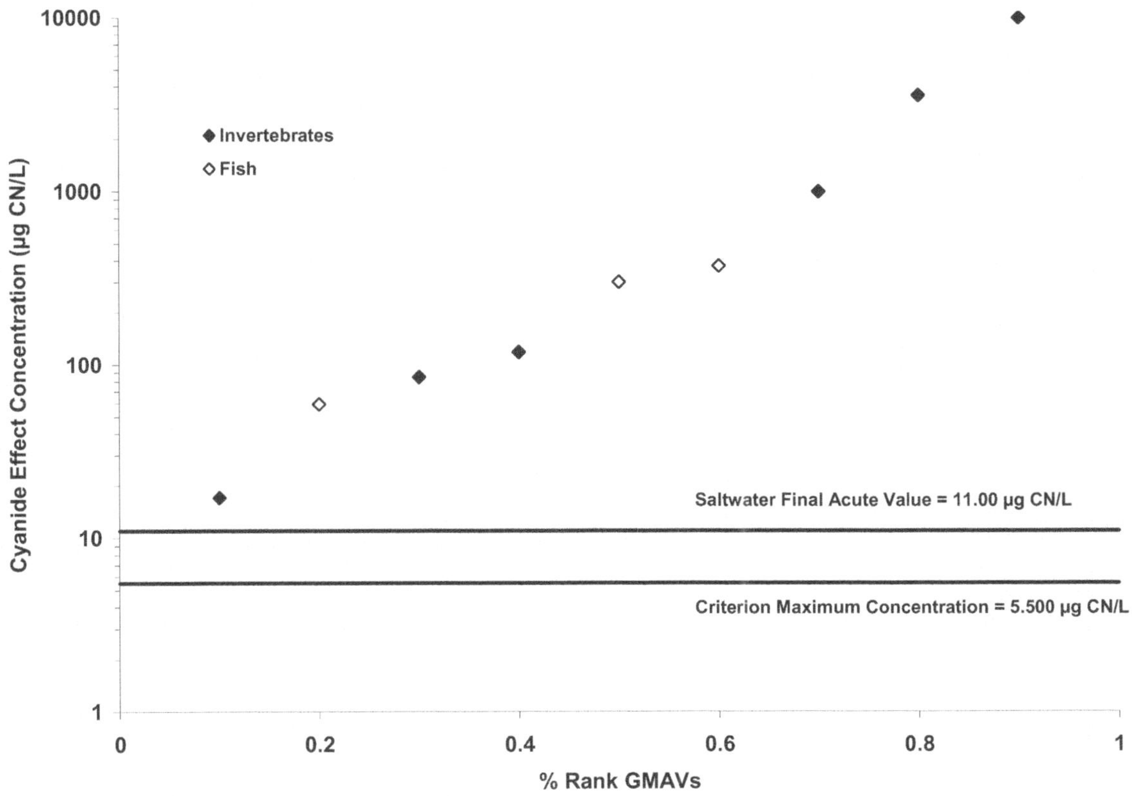

Figure 4-1. Ranked Summary of Cyanide GMAVs (Saltwater).

CHAPTER 5.0

TOXICITY TO AQUATIC PLANTS AND ALGAE

The mechanism of cyanide toxicity in plants is similar to that of animals, and is mediated through complexation with iron in the enzyme cytochrome oxidase (Köster, 2001). As with animals, this inhibits cellular respiration resulting in reduced ATP production. Without sufficient ATP levels, plants are incapable of conducting ion uptake and phloem transport can be inhibited, and if these functions are sufficiently impaired, the plant will eventually die (Köster, 2001).

In the 1984 AWQC (U.S. EPA, 1985a), cyanide toxicity data were identified for seven species of freshwater algae and vascular plants; no newer studies were identified (Table 4a). The most sensitive endpoint tested was growth inhibition, which occurred in the cyanobacterium *Microcystis aeruginosa* and the green algae *Scenedesmus quadricauda* exposed to 75 and 30 µg CN/L, respectively. In the diatom *Navicula seminuium*, 50% reductions in division rates were observed at cyanide concentrations ranging from 277-491 µg CN/L. In contrast, aquatic vascular plants were very resistant to cyanide. Potassium uptake was decreased in duckweed (*Lemna gibba*) exposed to 26,000 µg CN/L, and Eurasian watermilfoil (*Myriophyllum spicatum*) exposed to 22,400 µg CN/L had significantly lower root weights than control plants. The results of these studies suggest that cyanide appears to be more toxic to algae than vascular plants. The sensitivity of algae to cyanide is also within the range of that observed in many freshwater invertebrates, but less than the most acutely sensitive freshwater fish.

For saltwater species, cyanide toxicity data are available for several species of algae, but none are available for vascular plants (Table 4b). The most sensitive endpoints identified were associated with reproduction in the red algae *Champia parvula*. Growth of female *C. parvula* was significantly reduced when exposed to 11 µg CN/L, as was sexual reproduction. Two species of marine green algae were substantially less sensitive, with respiration in *Prototheca zopil* being significantly inhibited at 3,000 µg CN/L, and enzyme inhibition in *Chlorella* sp. Occurring at 30,000 µg CN/L (Table 4b).

In algal tests conducted since the 1984 AWQC (U.S. EPA, 1985a), the growth of the diatom *Nitzschia closterium* was found by Pablo et al. (1997b) to be significantly reduced when exposed to 10 µg CN/L for 72-hrs, with a 50% reduction in growth at 57 µg CN/L. Therefore, at least for the red algal *C. parvula* and the diatom *N. closterium*, saltwater algae can be even more sensitive to cyanide than the most acutely and chronically sensitive saltwater animal species. However, the recalculated chronic saltwater criteria for cyanide (Table 3c) would still be protective of growth and reproduction in even these most sensitive species.

CHAPTER 6.0

BIOACCUMULATION

There are no studies showing biomagnification of cyanide in the food chain. As summarized in U.S. EPA (1985a), cyanide does not bioaccumulate in aquatic organisms because it is readily metabolized. Therefore, the FCVs for cyanide (Table 3c) would not need to be adjusted to accommodate bioaccumulation of cyanide from dietary or food chain uptake.

CHAPTER 7.0

Toxicity to Freshwater Threatened and Endangered Species

7.1 Background

The current U.S. EPA Guidelines for deriving AWQC (U.S. EPA, 1985b) do not explicitly address protection of threatened and endangered (T&E) species. However, the U.S. EPA Guidelines were designed to ensure protection of economically, recreationally, or ecologically important species at levels of protection afforded to aquatic ecosystems under the Clean Water Act (CWA). Under the U.S. EPA's proposed revisions to the Guidelines, future derivation of AWQC will likely include an evaluation of whether AWQC are protective of T&E species at levels of protection afforded under the Endangered Species Act (ESA; U.S. EPA, 2003b). Furthermore, the U.S. EPA, U.S. Fish and Wildlife Service (FWS), and National Marine Fisheries Service (NMFS) have signed a Memorandum of Agreement (MOA) to better harmonize levels of protection afforded to aquatic organisms under the Clean Water Act relative to the Endangered Species Act (U.S. EPA, et al. 2001). Therefore, future updates to AWQC will likely require some level of analysis for protection of T&E species, even though no federal guidance currently exists for conducting such an assessment.

Under the current Guidelines (U.S. EPA, 1985b), AWQC are calculated as chemical concentrations that represent the 5th percentile of GMAVs based on available laboratory toxicity data, which are assumed to represent a random sample from a statistical population of GMAVs (Stephan, 2002). Although criteria calculated from the 5th percentile GMAV are intended to provide levels of protection required under the Clean Water Act (i.e., to protect biological, chemical, and ecological integrity of our nation's waters), by definition it is possible that some individuals of some taxa may not be protected by continuous exposure to a criterion concentration. It should be emphasized, however, that the 5th percentile applies to a hypothetical population of GMAVs and not to any particular field situation (Stephan, 2002).

In contrast, under the ESA, T&E species are to be protected at the level of the individual because the population is already impacted and may not be able to withstand further stress. Thus, if a T&E species is sensitive to a chemical near the criterion concentration, some individuals could theoretically be adversely affected. The less sensitive a T&E species is relative to the criterion concentration, the greater the likelihood that T&E individuals would not be affected, and so the criterion would be adequately protective even at the level of the individual organism. For cyanide in fresh waters, the level of protection is already assumed to be greater than 95% of the taxa since the acute and chronic AWQC were lowered to protect rainbow trout, the most sensitive species tested. To provide a more rigorous analysis, the rest of Section 7.0 evaluates whether the recalculated freshwater CMC and CCC of 23.27 and 4.817 µg/L, respectively, would likely be protective of T&E species at levels required under the ESA. Marine T&E species—

excluding mammalian and avian wildlife and sea turtles—are limited to the tidewater goby, Gulf and shortnose sturgeon, totoaba, smalltooth sawfish, and white abalone (NMFS, 2005). However, due to the lack of cyanide toxicity data for these species or reasonable surrogates, the protectiveness of the saltwater cyanide criteria to marine T&E species is not evaluated here. This represents an area requiring further research.

7.2 Overview of Approach

The following provides a brief summary of the approach used to evaluate whether the recalculated cyanide criteria are protective of freshwater T&E species (details of the analysis can be found in (DeForest et al., to be submitted for publication, 2007). In this T&E evaluation, two primary issues were considered. First was the sensitivity of T&E species relative to rainbow trout. Although rainbow trout (*Oncorhynchus mykiss*) is the most acutely sensitive species to cyanide based on toxicity studies considered acceptable according to U.S. EPA Guidelines (U.S. EPA, 1985b), other species, types of studies and endpoints were reviewed to determine whether any other empirical data suggest rainbow trout are not the most sensitive species to cyanide. In addition, the U.S. EPA Interspecies Correlation Estimations (ICE) model was used to estimate the sensitivity of a species to cyanide based on the known sensitivity of another species (Asfaw et al., 2003). The ICE model, therefore, provides another tool to evaluate the likelihood that other species (including T&Es and their surrogates) could be more sensitive to cyanide than rainbow trout.

The second issue to consider is the level of protection offered by any AWQC. For example, the acute cyanide criterion (i.e., CMC) is based on the species geometric mean of LC50 values for rainbow trout divided by two. Division of an LC50 by two is intended to estimate a low effect level that "is intended to result in a concentration that will not severely adversely affect too many of the organisms" (U.S. EPA, 1985b). To the extent that information was available, concentration-response curves for T&E surrogates were compared to the draft revised criteria to evaluate the level of protection offered by these criteria at the level of an individual organism.

Analysis of both main issues proceeded in a step-wise approach starting from lowest to highest empirical uncertainty, and culminating in an overall weight-of-evidence evaluation (Figure 7-1). The sensitivity of T&E species vs. AWQC concentrations was analyzed first using empirical toxicity data from T&E species themselves, and then using both intra-genus and intra-family laboratory surrogate species. Then, the toxicity of T&E species not tested or without appropriate surrogate tests was predicted using the ICE model both at the genus and family levels. Levels of protection for T&E species were then evaluated for both acute and chronic toxicity using tests from surrogate species for which appropriate dose-response data are available.

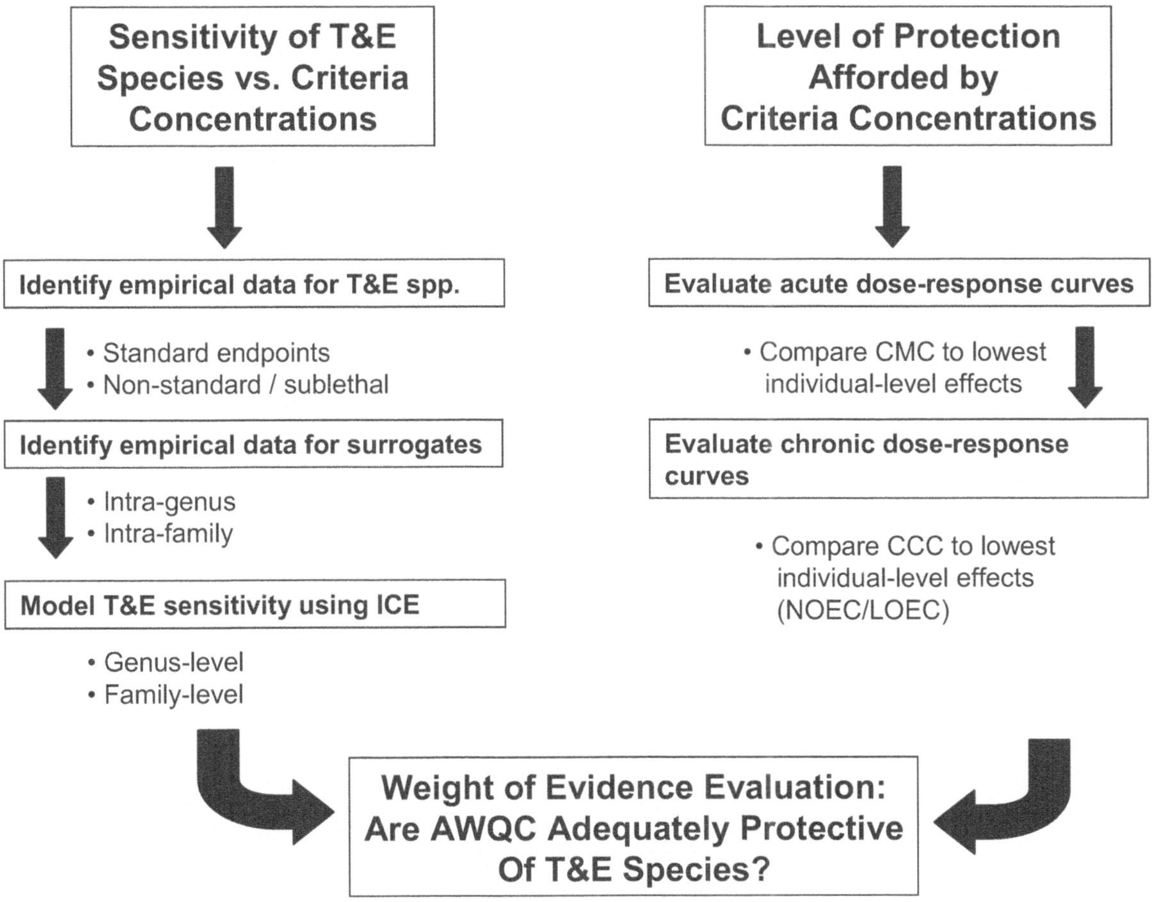

Figure 7-1. Multiple Lines of Evidence Approach for Evaluating the Protectiveness of Freshwater Cyanide AWQC to T&E Species.

7.3 Summary of Results

The primary results of the analysis of DeForest, et al. (to be submitted for publication, 2007) for protection of T&E species are summarized here.

Sensitivity of T&E Species Relative to Criteria Concentrations

Based on empirical studies with representatives of current T&E species, no cyanide toxicity values lower than the recalculated acute and chronic criteria were identified (Table 7-1). Further, the ICE model appears to be a reasonable tool for estimating cyanide LC50s for fish species lacking empirical data, and is more likely to conservatively underestimate LC50s. Figure 7-2 shows that a linear relationship exists between ICE-estimated LC50s for cyanide versus empirical SMAVs. Using the ICE model, cyanide LC50s could be estimated for nine T&E species of which five (the fountain darter *Etheostoma fonticola* and four salmonids) were less than the rainbow trout SMAV of 46.53 µg CN/L (DeForest et al., to be submitted for publication, 2007). If the lower confidence limit of the predicted LC50s is considered as a more conservative estimate of acute toxicity, the estimated cyanide LC50s for all nine T&E species would be less than the rainbow trout SMAV of 46.53 µg CN/L. In contrast, if the upper confidence limit of the predicted LC50s is considered, all would be greater than the rainbow trout SMAV.

Table 7-1. Empirical Cyanide Toxicity Data for T&E Species.

Species	Age	Duration	Endpoint	Conc. (µg CN/L)	Comment	Reference
Coho salmon *Oncorhynchus kisutch*	Juvenile	Up to 24 days	Growth	77	High disease and mortality at test initiation; salmon exposed to 19-77 µg CN/L in second half of study grew faster than controls.	(Leduc, 1966)
	Yearling	Up to 194 hrs.	Locomotor behavior (swimming time)	10	Reduced swimming time (3.8 min.) against a current compared to control fish (8.7 min.)	(Broderius, 1970)
Chinook salmon *Oncorhynchus tshawytscha*	Juvenile	2 mo.	Biomass	10 20	Increased biomass relative to control Reduced biomass relative to control	(Negilski, 1973)
Atlantic salmon *Salmo salar*[1]	6 mo. old	96 hrs.	LC50	90	Almost 2X the rainbow trout SMAV	(Tryland and Grande, 1983)

[1] T&E listing restricted to Gulf of Maine population.

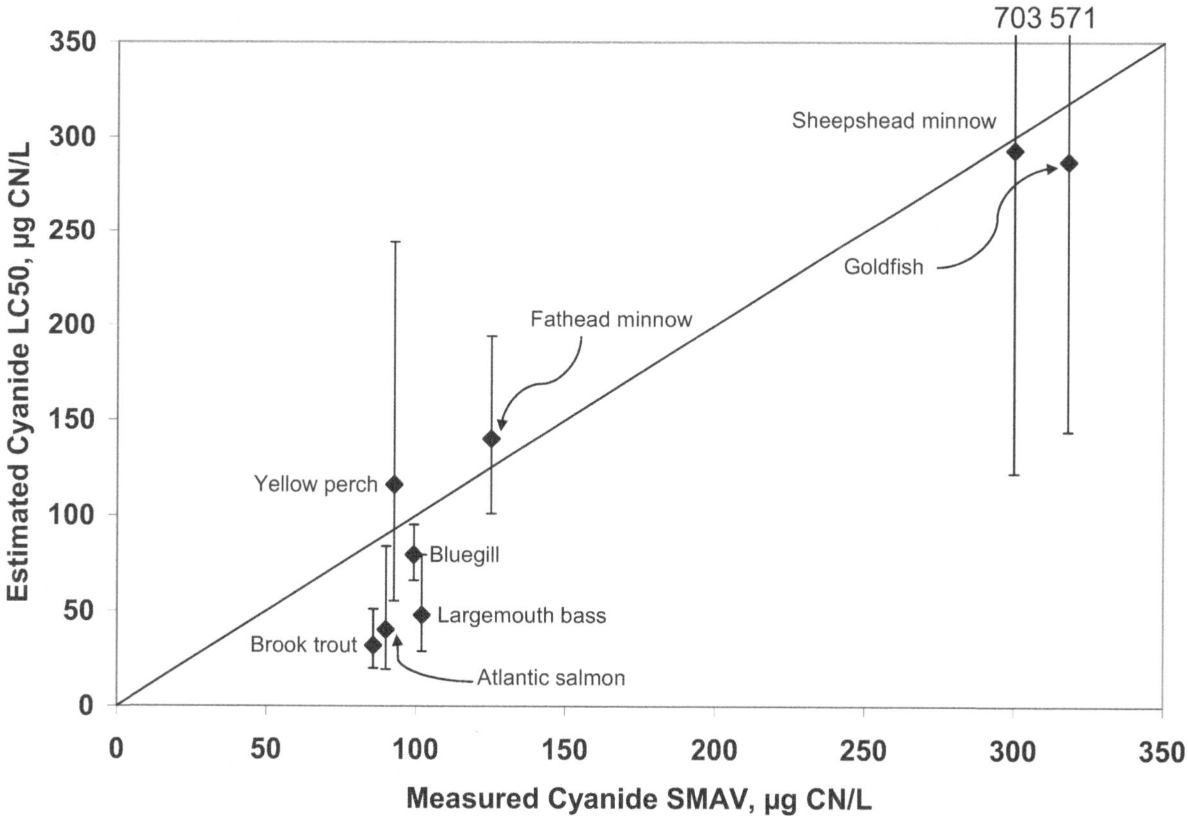

Figure 7-2. Relationship Between Measured Cyanide SMAVs and Estimated Cyanide LC50s using ICE (Solid Line Represents Perfect Agreement Between Measured and Estimated Values).

The ICE model was also used to estimate cyanide LC50s for species within families containing T&E species. This was possible for Families Cyprinidae, Cyprinodontidae, Ictaluridae, Percidae, and Salmonidae. The family mean LC50s calculated using the ICE model estimates were all greater than the rainbow trout-based SMAV of 46.53 µg CN/L, with the exception of Percidae, which had a family mean LC50 of 44 µg CN/L (DeForest et al., to be submitted for publication, 2007). This is consistent with the finding above that the fountain darter, a member of Family Percidae, may be more sensitive than rainbow trout.

All empirical and ICE-estimated 96-hr LC50s were combined to derive a species sensitivity distribution (SSD) for fish that could be used to estimate a possible distribution of cyanide LC50s for all fish (Figure 7-3). As shown, all measured and estimated LC50s are above the acute criterion of 23.27 µg CN/L.

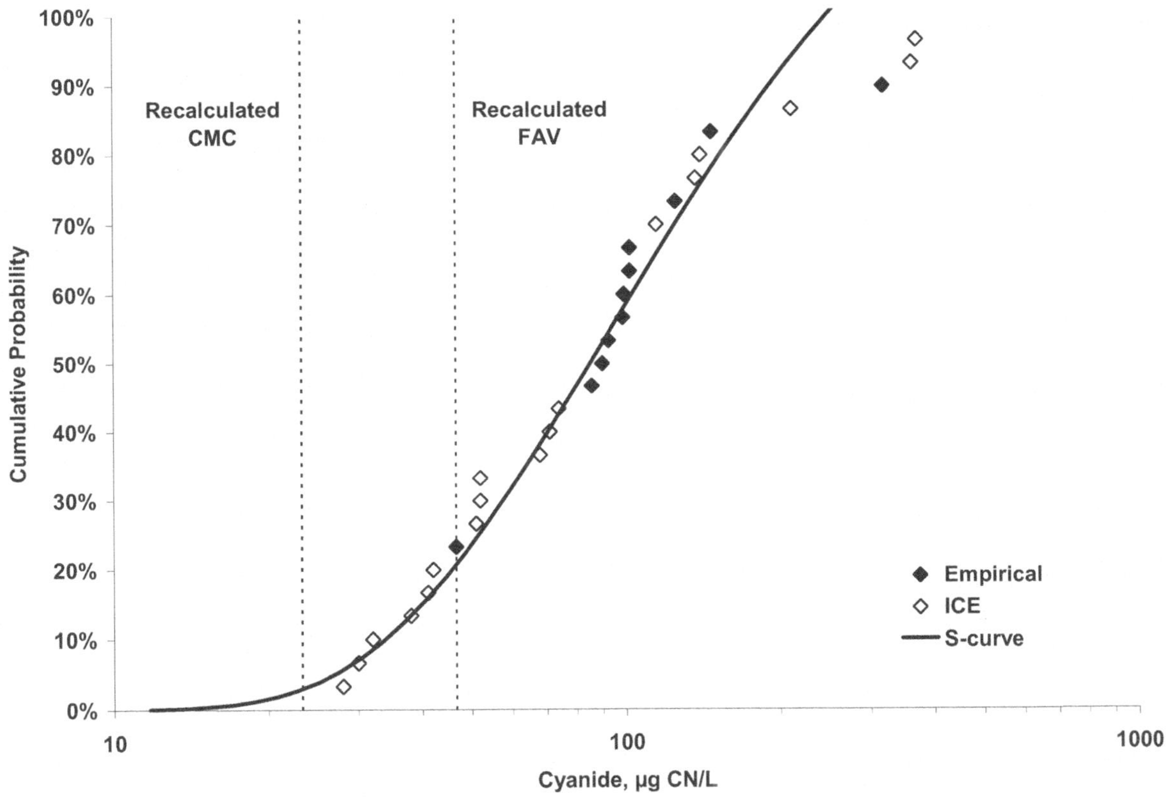

Figure 7-3. Species Sensitivity Distribution of Empirical and ICE-estimated Fish Species Mean Acute LC50 Values for Cyanide.

Level of Protection Afforded by Criteria Concentrations

In developing acute AWQC the FAV, which is based on LC50 values, is divided by two to estimate "a concentration that will not severely adversely affect too many of the organisms" (U.S. EPA, 1985b). The actual effect level estimated using this approach will vary depending on the steepness of the concentration-response curve for the chemical and species. This is because for steep dose-response curves, dividing an LC50 (or FAV based on LC50s) by two would represent a relatively low acute effect level. In contrast, for less steep dose-response curves, dividing an LC50 by two would not represent as low an effect level. For cyanide, the concentration-response curves for fish are quite steep (DeForest et al., to be submitted for publication, 2007), such that division of the LC50 by two often results in a cyanide concentrations less than the LC01. Based on cyanide toxicity data for rainbow trout (Kovacs and Leduc, 1982a; McGeachy and Leduc, 1988), the mean LC01:LC50 ratio was estimated to be 0.76. Multiplying the rainbow trout SMAV of 46.53 µg CN/L by 0.76 results in a cyanide concentration of 35.36 µg CN/L, which is still higher than the recalculated CMC. Thus, for cyanide, division of the rainbow trout SMAV by two appears to provide an additional margin of safety for fish species that are similarly or even slightly more sensitive to free cyanide than rainbow trout.

This LC01:LC50 ratio of 0.76 was then used to estimate LC01 values from the ICE-estimated LC50s for T&E species. The ICE model estimates an "expected" LC50 value, as well as 95% upper and lower confidence level estimates of the LC50. An assessment of whether the recalculated CMC is protective of the estimated LC01 values for T&E species is highly dependent on whether the estimated LC50 is based on the expected value or lower and upper confidence limits. Figure 7-4 show the ICE-estimated LC50 values, and corresponding lower and upper 95% confidence levels, for each T&E species, as well as the estimated LC01 values based on the LC01:LC50 ratio of 0.76. The recalculated CMC is also shown for comparison.

As shown in Figure 7-4, if the LC01 is estimated based on the 95% lower confidence limit LC50, all T&E species evaluated in ICE, with the exception of the bonytail chub, would appear to be under-protected by the recalculated CMC. If instead the LC01 is estimated based on the "expected" LC50 values, the resulting estimated LC01 values for only two T&E species are slightly lower than the recalculated CMC : fountain darter (estimated LC01 = 21.3 µg CN/L) and Apache trout (estimated LC01 = 22.8 µg CN/L). Finally, if the LC01 is estimated based on the 95% upper confidence limit LC50, all T&E species evaluated in ICE appear to be protected by the recalculated CMC (Figure 7-4).

Therefore, all but the most conservative ICE-based estimates of LC01 values suggest that T&E species would be adequately protected by the recalculated CMC, with the possible exception of the fountain darter and Apache trout based on the "expected" LC50 values. Even the most conservative estimate of the LC01, based on the ICE-predicted lower confidence limit, is only at most 45% lower than the recalculated CMC, which would mean dividing the CMC by two would be adequate to provide this additional level of protection if desired. However, the accuracy of the ICE-predicted lower confidence limits are difficult to independently confirm, and so LC01 values estimated from expected LC50 values is probably sufficiently conservative for this analysis.

Chronic Toxicity

Chronic toxicity data for cyanide are too limited to draw rigorous conclusions as to whether the chronic criterion is adequately protective of T&E species. If it is assumed that the ACR is appropriate and relevant across all acutely sensitive species, the chronic criterion should also be adequately protective of T&E species, with the exception of perhaps the fountain darter and Apache trout. Additional chronic studies, either with T&E species or other surrogate species, would be required to make more definitive conclusions on the protectiveness of the draft chronic criterion for T&E species.

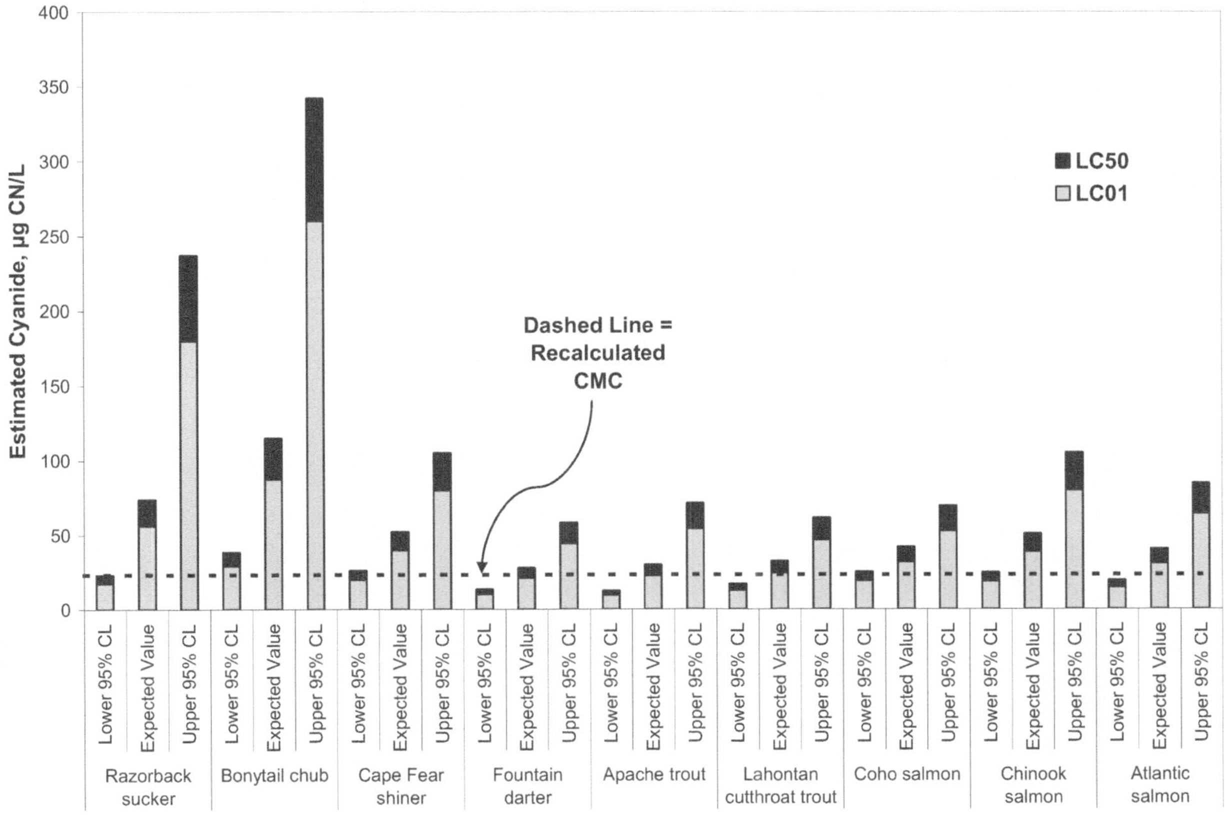

Figure 7-4. Comparison of Estimated Cyanide LC50 and LC01 Values for T&E Fish Species to the Recalculated Cyanide CMC.

Overall Weight-of Evidence

The empirical acute cyanide toxicity data for fish are very consistent, with SMAVs ranging from 46.53 to 147 µg/L for the nine most sensitive fish species (representing five families), suggesting that the sensitivity of fish to cyanide is not highly variable, and that the existing empirical data are likely to be representative of a variety of fish species. This suggests it is unlikely that a T&E fish species would be significantly more sensitive than rainbow trout, though if it were slightly more sensitive in terms of the LC50, division of the rainbow trout SMAV by two would be sufficiently conservative for estimating a negligible effect level. This is consistent with Dwyer et al. (2005) who evaluated the acute toxicity of five chemicals to rainbow trout and several T&E species. Based on the relative LC50s of these five chemicals to rainbow trout and T&E species, they recommended multiplying the geometric mean LC50 value for rainbow trout by 0.63 to estimate the LC50 for a listed species. Multiplying the rainbow trout cyanide SMAV of 46.53 µg CN/L by 0.63 results in an estimated LC50 of 29.31 µg CN/L for a listed species. If this estimated LC50 is multiplied by the mean LC01:LC50 ratio of 0.76 for rainbow trout, the estimated LC01 for a listed species would be 22.28 µg CN/L, which is very similar to the CMC of 23.27 µg CN/L.

Furthermore, use of the ICE model, which tends to provide conservatively low LC50 estimates for cyanide, along with estimation of very low effects levels (LC01), still only predicts that a small percentage of individuals in two T&E species (fountain darter and Apache trout) may be affected at the recalculated cyanide CMC. Given the conservative nature of this analysis, the cyanide CMC should be considered protective of most freshwater T&E species. However, removal of the most acutely sensitive coldwater species from the toxicity data set using the Recalculation Procedure (U.S. EPA, 1984, 1994) should still be done with caution to ensure they should not be retained as surrogates for untested T&E species. Further cyanide toxicity studies would also be helpful to confirm no- or low-acute effect levels for the most sensitive T&E species as predicted by ICE (i.e., fountain darter and Apache trout).

Uncertainties and Recommendations

The above conclusions must be considered in the context of the uncertainties associated with the above T&E analysis:

- Empirical acute cyanide toxicity data for T&E species are very limited
- Empirical acute cyanide toxicity data for T&E surrogates species are available for six of 14 families containing T&E fish species
- Empirical chronic cyanide toxicity for standard test organisms, not just T&E species, are lacking
- ICE is unable to predict the acute toxicity of cyanide to fish species in eight of the 14 families that contain T&E species
- ICE predictions may be affected by uncertainties in the interspecies correlation estimation models
- ICE predictions may be affected by variability and uncertainty of empirical LC50 data for surrogate species
- It was assumed that the slope of concentration-response curves for T&E species were the same as for rainbow trout

CHAPTER 8.0

TOXICITY TO AQUATIC DEPENDENT WILDLIFE

As discussed in Section 1, the U.S. EPA has been pursuing a more holistic approach in developing AWQC that accounts for protection of more receptor types via multiple exposure routes (U.S. EPA, 2003b). This section summarizes a review and analysis of cyanide toxicity to aquatic-dependent birds (e.g., sandpipers) and mammals (e.g., otters) recently conducted by Clark et al. (2006). Because sublethal cyanide doses are rapidly metabolized and excreted, cyanide does not bioaccumulate in animals. Consequently, the critical exposure pathway for cyanide poisoning in wildlife is likely via drinking water rather than dietary sources.

8.1 Methods for Compiling Avian and Mammalian Toxicity Data for Cyanide

Clark et al. (2006) reviewed the scientific literature to identify cyanide toxicity data for birds and mammals. Toxicity data were compiled based on the following exposure routes: dietary, drinking water, and injection. Reported data were screened according to the following criteria:

- Primary publications were used when possible, rather than review papers.
- The complete study design had to be detailed in the paper.
- Multiple doses had to be tested with evidence of a satisfactory dose-response relationship.
- Studies had to report either a lethal dose for 50% of a population (LD50), or no observable adverse effect level (NOAEL).

Most toxicity values were expressed as mg CN per kg body weight (mg CN / kg BW) in the original studies, or were converted to these units as necessary. When necessary, NOAELs were estimated from LD50s using an uncertainty factor of 100 (European Commission, 1996) and from lowest observable adverse effect levels (LOAELs) using an uncertainty factor of 10 (European Commission, 1996; U.S. EPA, 2003c).

8.2 Summary of Avian and Mammalian Toxicity Values for Cyanide

Cyanide toxicity values for birds and mammals, based on all exposure pathways, are plotted in Figure 8-1 (Clark et al., 2006). As shown, both birds and mammals exhibit a similar sensitivity to cyanide when exposed via drinking water—NOAELs for mammals range from 0.02 to 2.1 mg CN / kg BW and NOAELs for birds range from 0.01 to 0.8 mg CN / kg BW. Accordingly, Clark et al. (2006) suggested that a cyanide dose less than 0.01 mg CN / kg BW are protective of birds and mammals exposed to cyanide via drinking water and food. While this is orders of magnitude more conservative than the 1000 mg/kg threshold suggested by Eisler (1991), a threshold of 0.01 mg CN/kg BW protects from both water and dietary exposure pathways, and is consistent with sensitivity distributions for the most toxic simple cyanides regardless of exposure pathway.

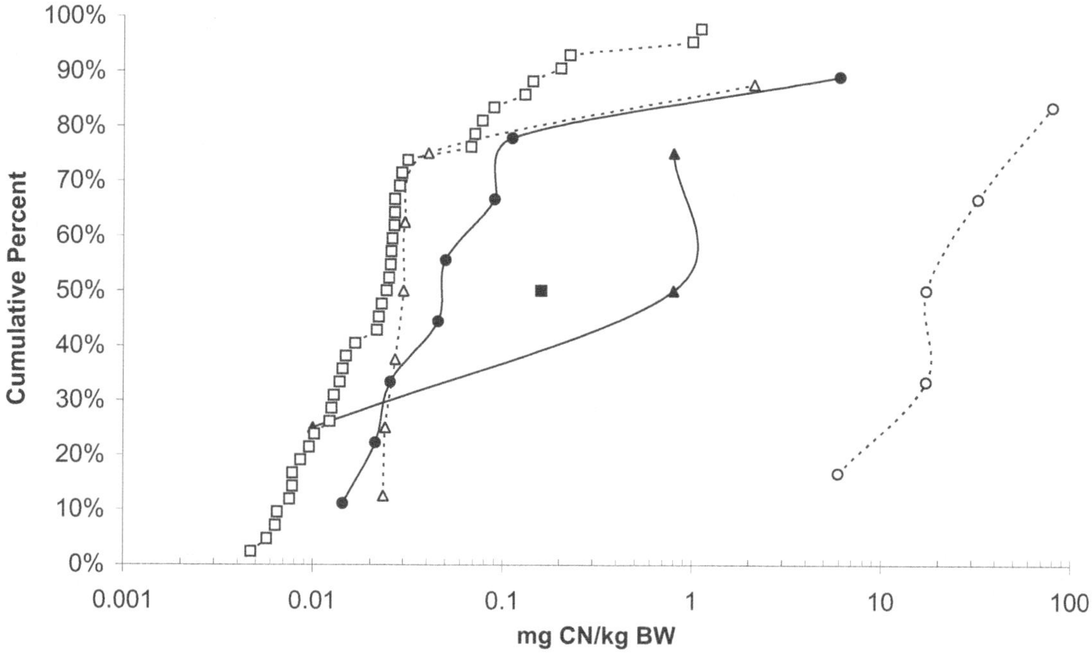

Figure 8-1. Toxicity of Cyanide to Birds and Mammals as a Function of Exposure Pathway, with Endpoints Normalized to Noaels Expressed as mg CN/kg BW (from Clark et al., 2006, with permission). Data are plotted using a cumulative distribution function of the ranked NOAELs. Data points corresponding to specific organisms and cyanide exposure pathways are denoted by (Δ) for mammals by drinking water, (○) for mammals by food, (□) for mammals by direct injection, (▲) for birds by drinking water, (●) for birds by food, and (■) for birds by direct injection.

8.3 Protectiveness of AWQC

Two different, but similar, approaches were used to evaluate whether the recalculated cyanide AWQC are adequately protective of aquatic-dependent analysis. Both analyses are based on exposure of wildlife to cyanide via drinking water, as this pathway is likely to be the most likely exposure route of birds and mammals to cyanide. The first approach was to develop a wildlife-based cyanide criterion estimated to be protective of 95% of wildlife species, which could then be compared to the recalculated acute and chronic freshwater criteria of 23 and 4.8 μg CN/L. The second approach was to estimate cyanide doses to wildlife assuming they ingested cyanide concentrations equal to the recalculated acute and chronic criteria, which could then be compared to the recommended safe dose of 0.01 mg CN/kg BW (Clark et al., 2006). Each analysis is presented separately below.

8.3.1 Approach #1: Cyanide Criterion for 95% of Wildlife Species

The proposed avian and mammalian wildlife toxicity threshold of 0.01 mg CN/kg BW can be translated into a surface water quality criterion based on acute exposures through use of several simple calculations. The strategy is to calculate water concentrations protecting a wide variety of water-dependent birds and mammal species, then calculate the water concentration that will protect 95% of these species. The latter is, in effect, analogous to an ambient water quality criterion for birds and mammals. The following explains the technical basis, details and results of this analysis.

Given information on representative wildlife receptor water ingestion rates and body weights, the first step was to estimate water cyanide concentrations that would result in the estimated safe dose of 0.01 mg CN/kg BW. Species-specific criteria were calculated using statistical methods similar to that of the FAV calculation (U.S. EPA, 1988) in addition to the Tier 2 methodology U.S. EPA introduced in the Great Lakes Initiative, and uses the following equation (U.S. EPA, 1993a):

$$\text{Species} - \text{Specific NOAEL (mg/L)} = \frac{(\text{NOAEL x BW})}{\text{WIR}} \qquad \textbf{(8-1)}$$

where: NOAEL = no observed adverse effect level (0.01 mg CN per kg BW per day)
WIR = water ingestion rate (L/day)
BW = body weight (kg)

The species selected for this analysis were identified to represent a broad cross-section of water-dependent birds and mammals representing different trophic levels and taxa:

- Small insectivorous mammals (water shrew)
- Medium herbivorous mammals (muskrat and beaver)
- Medium omnivorous mammals (raccoon)
- Large herbivorous mammals (deer and moose)
- Predacious mammals (mink)
- Small insectivorous bird (swallow)
- Small piscivorous bird (kingfisher)
- Small shorebird (spotted sandpiper)
- Small marsh bird (snipe)
- Medium herbivorous bird (mallard)
- Medium piscivorous bird (blue heron)
- Large herbivorous bird (goose)
- Large predacious bird (bald eagle)

Body weights for these receptors were identified from the scientific literature and drinking water ingestion rates were generally estimated from body weight using allometric equations. Allometric equations are empirically derived estimates of water uptake that are based on the relationship between body weight and water uptake for many species (U.S. EPA, 1993b). The allometric equations for estimating drinking water intake rates for birds and mammals are as follows:

Birds: $$WIR = 0.059 \times BW^{0.67} \quad \textbf{(8-2)}$$

Mammals: $$WIR = 0.099 \times BW^{0.90} \quad \textbf{(8-3)}$$

where: WIR = water ingestion rate (L/day)
BW = body weight (kg)

The body weights and water ingestion rates for the wildlife receptors are provided in Table 8-1. Using this information in Equation 8-1, the equivalent free cyanide water concentrations were estimated for each species (Table 8-2). The species were then ranked from low to high in terms of water-based toxicity thresholds and then plotted as a cumulative probability distribution.

Table 8-1. Body Weights and Water Uptake Rates of Representative Avian and Mammalian Wildlife Species.

Receptor	Body Weight (kg)	Water Ingestion Rate (L/day)	References
Cliff swallow	0.0146	0.0035	(Dunning, 1993; U.S. EPA 1993a)
Spotted sandpiper	0.042	0.007	(U.S. EPA, 1993a, 1993b)
Common snipe	0.122	0.0144	(Dunning, 1993; U.S. EPA, 1993a)
Belted kingfisher	0.147	0.016	(U.S. EPA, 1993a, 1993b)
Mallard	1.159	0.0649	(U.S. EPA, 1993a, 1993b)
Great blue heron	2.3	0.1	(U.S. EPA, 1993a, 1993b)
Canada goose	2.99	0.1226	(U.S. EPA, 1993a, 1993b)
Bald eagle	3.67	0.13	(U.S. EPA, 1993a, 1993b)
Water shrew	0.0125	0.003	(Conway, 1952; U.S. EPA, 1993b)
Mink	0.8	0.081	(U.S. EPA, 1995)
Muskrat	1.15	0.11	(U.S. EPA, 1993a, 1993b)
Raccoon	5.42	0.45	(U.S. EPA, 1993a, 1993b)
Beaver	19.31	1.42	(U.S. EPA, 1993a, 2003c)
White tailed deer	69.42	4.5	(U.S. EPA, 1993a, 2003c)
Moose	539.2	28.5	(U.S. EPA, 1993a; Rause and Gasaway, 1994)

Table 8-2. Calculated Equivalent Free Cyanide Water Concentration with Ranking.

Receptor	Toxicity Threshold Concentration (mg CN/L)	Rank (%)
Cliff swallow	0.042	6
Water shrew	0.045	13
Spotted sandpiper	0.060	19
Common snipe	0.085	25
Belted kingfisher	0.090	31
Mink	0.099	38
Muskrat	0.102	44
Raccoon	0.119	50
Beaver	0.136	56
Bald eagle	0.139	63
White tailed deer	0.154	69
Mallard	0.179	75
Moose	0.189	81
Heron	0.222	88
Canada goose	0.244	94

The last step was to estimate the toxicity threshold that would protect 95% of the wildlife species shown in Table 8-2, which were assumed to represent aquatic-dependent wildlife species in general. To estimate a concentration protecting 95% of the wildlife species, various distributions were fit to the data in Table 8-2 using the software program BestFit (Palisade Corporation, 1996), and the best-fitting distributions are shown in Figure 8-2. The cyanide concentrations in drinking water estimated to protect 95% of the species from acute cyanide toxicity were then estimated from the best fitting distributions (Figure 8-2). Although each of the distributions fit the data differently, especially the lower tails, the 95% protection levels were very uniform, averaging 0.04 ± 0.006 mg CN/L. The geometric mean 95% protection level is 0.039 mg CN/L and the lowest estimate is 0.032 mg CN/L. Thus, the recalculated aquatic life-based CMC of 23 μg CN/L would be protective of even the lowest estimated 95% protection level of 0.032 mg CN/L (32 μg CN/L).

8.3.2 Approach #2: Cyanide Criterion Based on Safe Dose

In the second approach for evaluating whether the recalculated AWQC are protective of aquatic-dependent wildlife, the allometric equations for birds and mammals used above (Equations 8-2 and 8-3) were used to estimate water ingestion rates across a large range of avian and mammalian body weights. These body weight-water ingestion rate combinations were then multiplied by the recalculated acute and chronic criteria of 23 and 4.8 μg CN/L to estimate cyanide doses (mg CN/kg BW) to a hypothetical range of wildlife species:

$$\text{Estimated Dose} = (\text{Recalculated Criterion} \times \text{WIR}) / \text{BW} \quad \textbf{(8-4)}$$

Where:			
	Recalculated Criterion	=	0.023 or 0.0045 mg CN/L (23 or 4.5 μg CN/L)
	WIR	=	water ingestion rate (L/day)
	BW	=	body weight (kg)

These estimated cyanide doses can than be compared to Clark et al. (2006) suggested safe cyanide dose of 0.01 mg CN / kg BW. Figure 8-3 provides the results of this evaluation for birds (the results for mammals were virtually identical). As shown, if birds were exposed to the acute criterion, they would have to consume drinking water at a rate greater than 40 percent of their body weight per day to reach the suggested safe dose of 0.01 mg CN / kg BW. For comparison, the water intake rates for bird species profiles provided by U.S. EPA (1993b) tend to range from approximately 5 to 30 percent of the birds body weight per day. Consequently, it appears to be unlikely that a bird, even if exposed continuously to a cyanide concentration equivalent to the recalculated acute criterion of 23 µg CN/L, would reach the suggested safe dose of 0.01 µg CN / kg BW. If exposed continuously to the recalculated chronic criterion of 4.8 µg CN/L, the likelihood is even lower. Accordingly, based on this analysis and the analysis described under Approach #1 above, the recalculated acute and chronic cyanide criteria would be protective of aquatic-dependent birds and mammals.

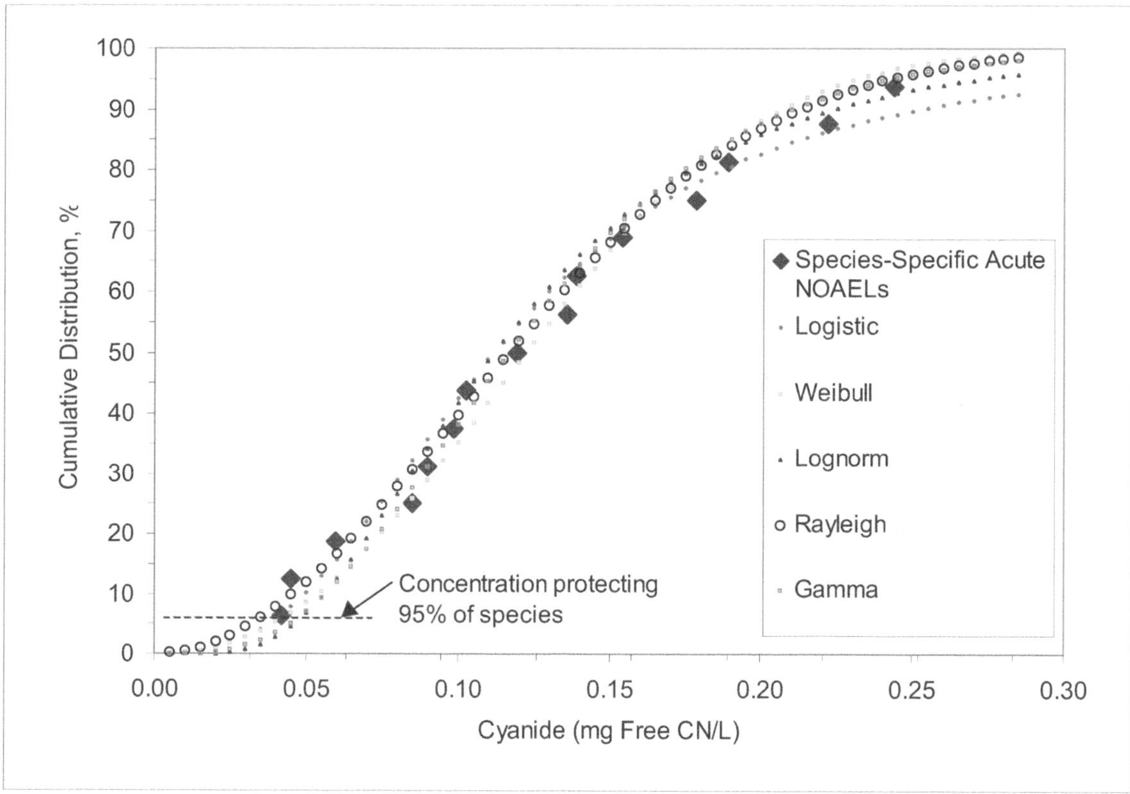

Figure 8-2. Best Fitting Cumulative Probability Distributions to the Estimated Cyanide Noaels for Wildlife.

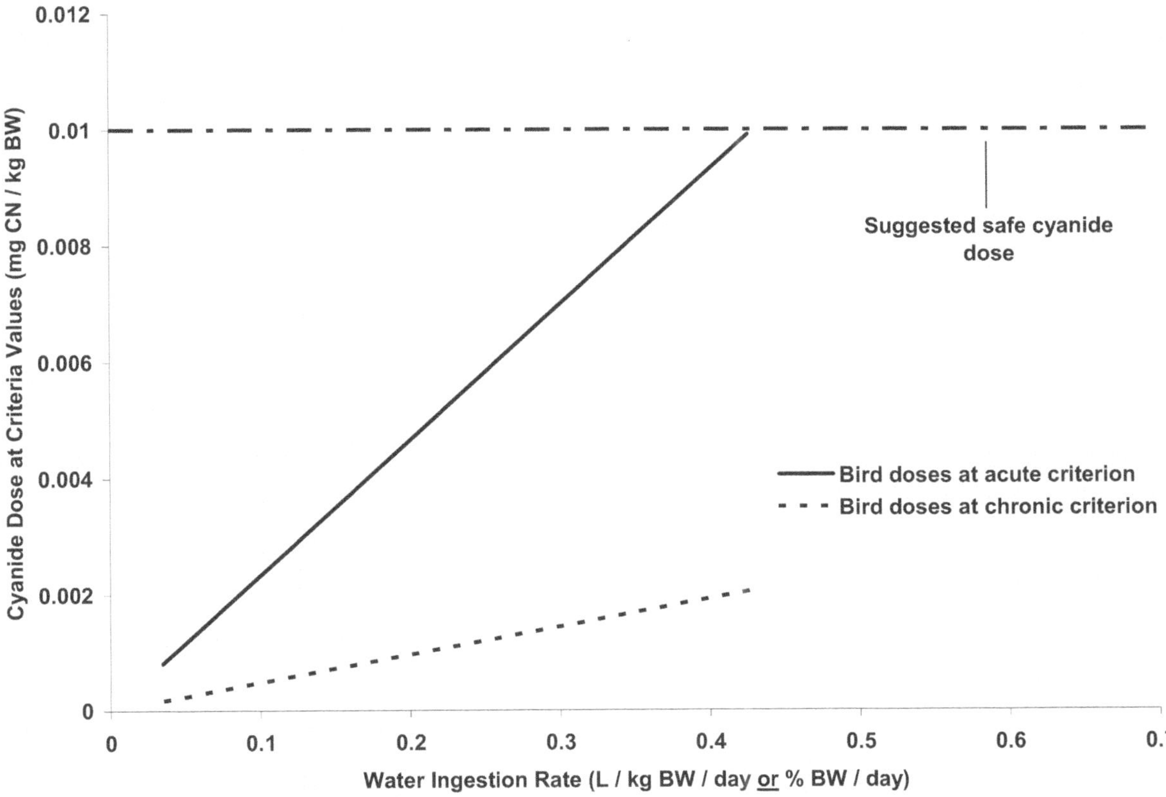

Figure 8-3. Estimated Cyanide Doses to Birds Exposed to Acute and Chronic Criteria Concentrations In Drinking Water.

CHAPTER 9.0

TOXICITY TO BENTHIC (SEDIMENT) ORGANISMS

In terms of water exposures to free cyanide, benthic invertebrates appear to be much less sensitive to free cyanide than most fish species. Of the 27 GMAVs for freshwater species, 12 are for benthic species (or species with benthic life stages) and these fall within GMAV ranks of 14 to 27. Consequently, the recalculated AWQC should be protective of benthic invertebrates given that they appear to be less acutely sensitive to free cyanide than are fish or other invertebrates. However, empirical data on the toxicity of free cyanide to benthic invertebrates tested via sediment exposures are very limited. The Equilibrium Partitioning (EqP) sediment guidelines (U.S. EPA, 2003d), however, provide a means for estimating bulk sediment concentrations of free cyanide that are protective. The EqP approach assumes that bioavailable forms of a substance in the sediment porewater mediate toxicity to benthic organisms. Free cyanide in water overlying sediment will eventually contaminate the sediment, partly because soluble cyanide can diffuse into the sediment porewater, and partly because free cyanide can partition to sediment particles. Either form of cyanide in sediments will contribute to the bulk sediment concentration. The bulk sediment concentration can be estimated, therefore, by assuming the sediment porewater is in diffusive equilibrium with the overlying water (i.e., overlying water and porewater constituent concentrations are the same) and then estimating the additional contribution of partitioning on sediment surfaces to the bulk sediment concentration. There is potential for free cyanide that diffuses into sediments to complex with metals such as iron, thereby reducing the toxicity of cyanide (Gensemer et al., 2006). For the purposes of conservative evaluation of risk to benthic organisms, the potential for metal-cyanide complexation thus can be ignored.

This mode of sediment contamination (i.e., from small concentrations of free cyanide in the overlying water near the criterion value) is in distinct contrast to heavily contaminated sites where cyanide concentrations in soils and sediments can become quite high due to the disposal or formation of persistent cyanide precipitates with iron and other metals. It is clear that such heavily contaminated sites could possibly result in free cyanide concentrations well in excess of the recalculated AWQC for cyanide. Industrially contaminated sediments are not relevant to the issue of whether the free cyanide AWQC is protective of benthic organisms, and thus this scenario can be excluded from consideration.

Partitioning of free cyanide to sediments was studied by Higgins and Dzombak (2006) for several mineral phases as well as natural sediments from a freshwater lake. Free cyanide adsorbs very weakly to mineral surfaces, but can bind to sediment organic matter enough to produce measurable bulk sediment concentrations. The preferential binding of free cyanide to organic matter was confirmed by comparing the carbon-normalized binding of free cyanide in sediments to binding in activated carbon. The overall level of binding in natural sediments was still fairly low, even considering the contribution of sediment organic matter, given an organic-carbon-normalized distribution coefficient K_{oc} ($=C_s/C_w f_{oc}$) value of approximately 12.9 L/g_s under freshwater conditions. The weak binding of free cyanide to natural sediments makes it unlikely that free cyanide will accumulate to any appreciable extent. Furthermore, there is ample evidence that low levels of free cyanide are readily degraded by microbial processes and do not persist in biologically active soils and sediments (Ebbs, et al., 2006). If it is assumed that benthic organisms are not inordinately more sensitive to cyanide than the water column organisms that drive the basis for the recalculated freshwater and saltwater criteria (the available cyanide toxicity data do not suggest that they would be) and given that bioavailable forms of cyanide are not expected to accumulate appreciably in sediment, a sediment-based cyanide criterion does not appear warranted. Cyanide should be effectively regulated based on water column-based criteria and concentrations.

CHAPTER 10.0

OTHER DATA

Toxicity studies that do not meet the requirements for inclusion in Tables 1-4, but otherwise provide useful information on the effects of cyanide to aquatic organisms typically are included as "other data" (U.S. EPA, 1985b). These other toxicity data for freshwater and saltwater species are provided in Tables 5a and 5b, respectively. A number of studies have reported various sublethal effects in salmonids at cyanide concentrations near 10 µg CN/L (see also Gensemer et al., 2006), but these effects were observed at concentrations greater than the recalculated FCV of 4.457 µg CN/L (and corresponding no-effect concentrations are generally unavailable). A summary of these studies is provided below.

Exposure of rainbow trout (*O. mykiss*) to a cyanide concentration of 9.6 µg CN/L has been shown to result in reduced weight gain, increased respiration rate, liver damage, and a higher relative body water content (Dixon and Leduc, 1981), altered blood chloride and osmolarity (Leduc and Chan, 1975), abnormal oocyte development (Lesniak, 1977; Lesniak and Ruby,1982), reduced production of dividing spermatogonia (Ruby, 1979), and reduced wet weight gain in fish forced to swim against a 12 cm/sec current (McCracken and Leduc, 1980). Exposure of Atlantic salmon (*S. salar*) to 9.6 µg CN/L resulted in abnormal embryo and larval development (Leduc, 1978). Furthermore, as discussed in Section 7, studies with coho (*O. kisutch*) and chinook (*O. tshawytscha*) salmon suggest that a cyanide concentration as low as 10 µg CN/L reduces the ability of yearling coho salmon to swim against a current (Broderius, 1970) and a cyanide concentration of 20 µg CN/L may be sufficient to reduce chinook salmon biomass (Negilski, 1973). All other toxicity data for saltwater species are well above the recalculated saltwater CMC and FCV.

Overall, none of these other toxicity values for freshwater are lower than the recalculated cyanide CCC derived in Sections 3.0 and 4.0. Therefore, the CCC does not need to be lowered further to protect against these other sublethal or non-standard chronic endpoints.

CHAPTER 11.0

UNUSED DATA

Based on the U.S. EPA Guidelines (U.S. EPA, 1985b), the following studies were considered unacceptable for deriving AWQC for the following reasons, and are thus classified as unused data.

Studies Were Conducted with Species That Are Not Resident in North America

(Abram, 1964)
(Alazemi et al., 1996)
(Brockway, 1963)
(Costa, 1966)
(Jones and Hoegh-Guldberg, 1999)
(Jones and Steven, 1997)
(Jones et al., 1999)
(Lomte and Jadhav, 1982)
(Pablo et al., 1997[a]; Pablo et al., 1997c)
(Rippon et al., 1992)
(Sanoli and Kanabur, 2001)
(Sarkar, 1990)
(Solbe et al., 1985)
(Woker and Wuhrmann, 1950)

Cyanide Was a Component of a Complex Cyanide or Effluent

(Doudoroff, 1976)
(Lloyd and Jordan, 1964)
(Shelford, 1917)

Data Not Used Because Results Were Only Presented Graphically

(Downing, 1954)
(Renn, 1955)
(Smith and Heath, 1979)

Studies Were Conducted Using Inadequate Dilution Water, Using Inappropriate Test Methods, or Too Little Information on Test Conditions

(Bridges, 1958)
(Bringmann and Kuhn, 1982)
(Buikema, 1977)
(Costa, 1965b, 1965c, 1965a)
(Ishio, 1965)
(Jones, 1941)
(Lewis and Tarrant, 1960)
(Lipschuetz and Cooper, 1955)
(Lund, 1918)
(Moore and Kin, 1968)
(Washburn, 1948)
(Whittingham, 1952)
(Summerfelt and Lewis, 1967)

CHAPTER 12.0

SUMMARY AND AWQC RECALCULATIONS

The AWQC for (free) cyanide originally developed in 1984 (U.S. EPA, 1985a) have been re-evaluated with consideration of new toxicity data published since that time and in the framework of current U.S. EPA protocols for AWQC determination. The procedures described in the *Guidelines for Deriving Numerical National Water Quality Criteria for the Protection of Aquatic Organisms and Their Uses*" (U.S. EPA, 1985b) indicate that, except possibly where a locally important species is very sensitive, freshwater aquatic organisms and their uses should not be affected unacceptably if the four-day average concentration of cyanide does not exceed 4.8 μg CN/L more than once every three years on the average and if the one-hour average concentration does not exceed 23 μg CN/L more than once every three years on average. If coldwater species (e.g., salmonid fish) are not associated with a waterbody, freshwater organisms and their uses should not be affected unacceptably if the four-day average concentration of cyanide does not exceed 9.8 μg CN/L more than once every three years on the average, and if the one-hour average concentration of cyanide does not exceed 47 μg CN/L more than once every three years on the average.

The procedures described in the *Guidelines for Deriving Numerical National Water Quality Criteria for the Protection of Aquatic Organisms and Their Uses*" (U.S. EPA, 1985b) indicate that, except possibly where a locally important species is very sensitive, saltwater aquatic organisms and their uses should not be affected unacceptably if the four-day average concentration of cyanide does not exceed 1.1 μg CN/L more than once every three years on the average, and if the one-hour average concentration does not exceed 5.5 μg CN/L more than once every three years on average.

The analyses and procedures outlined elsewhere in this report indicate that, except where a locally important species is very sensitive, that most freshwater species listed as federally threatened or endangered under the Endangered Species Act and their uses should not be affected unacceptably if the four-day average concentration of cyanide does not exceed 4.8 μg CN/L more than once every three years on the average, and if the one-hour average concentration does not exceed 23 μg CN/L more than once every three years on average. The overall weight-of-evidence indicated that these recalculated AWQC should be protective of T&E species even at very conservative levels of protection, although the ICE model predicts that the fountain darter (*Etheostoma fonticola*), Apache trout (*Oncorhynchus apache*), and sensitive members of the fish family Percidae may not be adequately protected by the acute criterion at the level of the individual organism. In such cases, dividing the CMC or CCC by a factor of 2 should provide an adequately conservative level of additional protection for species of special concern.

The analyses and procedures outlined in this report indicate that, except where a locally important species is very sensitive, aquatic-dependent wildlife (i.e., avian and mammalian wildlife) and their uses should not be affected unacceptably if the four-day average concentration of cyanide does not exceed 4.8 μg CN/L more than once every three years on the average, and if the one-hour average concentration does not exceed 23 µg CN/L more than once every three years on average. Because cyanide does not bioaccumulate, the primary exposure pathway is via drinking water, particularly during acute spill events. Allometric models predict that reasonable exposure scenarios at or below the recalculated CCC would not cause birds or mammals to exceed a suggested safe dose of 0.01 µg CN / kg BW.

Due to the relatively weak partitioning of free cyanide to sediments as well as its biodegradability, the recalculated AWQC are likely to be adequately protective of benthic organisms. Thus, freshwater benthic (sediment) organisms would not be affected unacceptably if the four-day average concentration of cyanide does not exceed 4.8 µg CN/L more than once every three years on the average, and if the one-hour average concentration does not exceed 23 µg CN/L more than once every three years on average. Saltwater benthic organisms should not be affected unacceptably if the four-day average concentration of cyanide does not exceed 1.1 µg CN/L more than once every three years on the average, and if the one-hour average concentration does not exceed 5.5 µg CN/L more than once every three years on average.

With the development of more reliable analytical methods for measuring free cyanide, the U.S. EPA currently recommends measuring free cyanide (U.S. EPA, 1985a) to determine compliance with AWQC, as the criteria were developed on this basis. However, when the current AWQC for cyanide was published in 1984, no approved U.S. EPA methods for measurement of free cyanide were available. Therefore, the U.S. EPA recommended measuring total cyanide and recognized that the criteria may be overly protective when total cyanide is analyzed to determine compliance. Given that approved methods now exist for measuring free cyanide, such as microdiffusion (ASTM, 1998b), it is recommended that free cyanide now be used to determine compliance with cyanide AWQC. Alternatively, measurement of weak acid dissociable (WAD) cyanide (method 4500 CN- I; American Public Health Association, 1998) provides a reasonable estimate of free cyanide that errs on the side of a conservative estimate of free cyanide.

APPENDIX A

TOXICITY DATA TABLES TO SUPPORT CRITERIA CALCULATIONS

Table 1a. Acute Toxicity of Cyanide to Freshwater Animals.

Freshwater Species	Method	LC50 or EC50 (µg CN/L)	SMAV (µg CN/L)	Analytical Method Classification	Reference
Asellus communis (isopod)	FT, M	2,326	2,326	2	(Oseid and Smith, 1979)
Asellus intermedius (isopod)	S, U	1,700	1,700	6	(Ewell et al., 1986)
Carassius auratus (goldfish)	FT, M	318	318	2	(Cardwell et al., 1976)
Corixa sp. (water boatman)	S, U	247	250	6	(Sarkar, 1990)
Corixa sp. (water boatman)	S, U	251		6	(Sarkar, 1990)
Corixa sp. (water boatman)	S, U	252		6	(Sarkar, 1990)
Cyclops viridis (copepod)	S, U	78	130	6	(Sarkar, 1990)
Cyclops viridis (copepod)	S, U	167		6	(Sarkar, 1990)
Cyclops viridis (copepod)	S, U	169		6	(Sarkar, 1990)
Daphnia magna (cladoceran)	S, U	1800*	120	6	(Anderson ,1946)
Daphnia magna (cladoceran)	S, U	160		6	(Dowden and Bennett, 1965)
Daphnia magna (cladoceran)	S, U	90		6	(Ewell et al., 1986)
Daphnia pulex (cladoceran)	S, M	110	96	6**	(Cairns et al., 1978)
Daphnia pulex (cladoceran)	S, U	83		6	(Lee, 1976)
Daphnia spp. (cladoceran)	S, U	80	132	6	(Sarkar, 1990)
Daphnia spp. (cladoceran)	S, U	169		6	(Sarkar, 1990)
Daphnia spp. (cladoceran)	S, U	172		6	(Sarkar, 1990)
Diaptomus sp. (copepod)	S, U	82	133	6	(Sarkar, 1990)
Diaptomus sp. (copepod)	S, U	166		6	(Sarkar, 1990)
Diaptomus sp. (copepod)	S, U	172		6	(Sarkar, 1990)
Dugesia tigrina (planarian)	S, U	2,100	2,100	6	(Ewell,et al., 1986)
Dytiscus sp. (aquatic beetle)	S, U	246	251	6	(Sarkar, 1990)
Dytiscus sp. (aquatic beetle)	S, U	249		6	(Sarkar, 1990)
Dytiscus sp. (aquatic beetle)	S, U	259		6	(Sarkar, 1990)
Gammarus fasciatus (amphipod)	S, U	900	900	6	(Ewell et al., 1986)
Gammarus pseudolimnaeus (amphipod)	FT, M	167	167	2	(Oseid and Smith, 1979)
Gasterosteus aculeatus (three-spine stickleback)	FT, M	98.8	98.8	1	(Parametrix, 2005)
Helisoma trivolvis (snail)	S, U	>50,000	>50,000	6	(Ewell et al., 1986)

Table 1a. Acute Toxicity of Cyanide to Freshwater Animals.

Freshwater Species	Method	LC50 or EC50 (µg CN/L)	SMAV (µg CN/L)	Analytical Method Classification	Reference
Lepomis macrochirus (bluegill)	S, U	180*	99	6	(Cairns and Scheier, 1958, 1968; Patrick, et al., 1968)
Lepomis macrochirus (bluegill)	S, M	220*		-	(Cairns and Scheier, 1959)
Lepomis macrochirus (bluegill)	S, M	180*		-	(Cairns and Scheier, 1959)
Lepomis macrochirus (bluegill)	S, M	230*		-	(Cairns and Scheier, 1959)
Lepomis macrochirus (bluegill)	S, M	160*		4	(Cairns and Scheier, 1963)
Lepomis macrochirus (bluegill)	S, M	150*		-	(Henderson et al., 1961)
Lepomis macrochirus (bluegill)	FT, M	364*		2	(Smith et al., 1978)
Lepomis macrochirus (bluegill)	FT, M	232*		2	(Smith et al., 1978)
Lepomis macrochirus (bluegill)	FT, M	279*		2	(Smith et al., 1978)
Lepomis macrochirus (bluegill)	FT, M	273*		2	(Smith et al., 1978)
Lepomis macrochirus (bluegill)	FT, M	81		2	(Smith et al., 1978)
Lepomis macrochirus (bluegill)	FT, M	85.7		2	(Smith et al., 1978)
Lepomis macrochirus (bluegill)	FT, M	74		2	(Smith et al., 1978)
Lepomis macrochirus (bluegill)	FT, M	100		2	(Smith et al., 1978)
Lepomis macrochirus (bluegill)	FT, M	107		2	(Smith et al., 1978)
Lepomis macrochirus (bluegill)	FT, M	99		2	(Smith et al., 1978)
Lepomis macrochirus (bluegill)	FT, M	113		2	(Smith et al., 1978)
Lepomis macrochirus (bluegill)	FT, M	121		2	(Smith et al., 1978)
Lepomis macrochirus (bluegill)	FT, M	126		2	(Smith et al., 1978)
Lumbriculus variegatus (oligochaete)	S, U	11,000	11,000	6	(Ewell et al., 1986)
Micropterus salmoides (largemouth bass)	FT, M	102	102	2	(Smith et al., 1978)
Nepa sp. (water scorpion)	S, U	241	273	6	(Sarkar, 1990)
Nepa sp. (water scorpion)	S, U	289		6	(Sarkar, 1990)
Nepa sp. (water scorpion)	S, U	293		6	(Sarkar, 1990)
Oncorhynchus mykiss (rainbow trout)	S, U	90*	47	6	(Bills, et al., 1977)
Oncorhynchus mykiss (rainbow trout)	FT, M	27		2	(Kovacs, 1979)
Oncorhynchus mykiss (rainbow trout)	FT, M	40		2	(Kovacs, 1979)
Oncorhynchus mykiss (rainbow trout)	FT, M	65		2	(Kovacs, 1979)
Oncorhynchus mykiss (rainbow trout)	S, U	46.3*		6	(Marking et al., 1984)
Oncorhynchus mykiss (rainbow trout)	S, U	52.1*		6	(Marking et al., 1984)
Oncorhynchus mykiss (rainbow trout)	S, U	54.1*		6	(Marking et al., 1984)
Oncorhynchus mykiss (rainbow trout)	S, U	62.1*		6	(Marking et al., 1984)
Oncorhynchus mykiss (rainbow trout)	S, U	74.8*		6	(Marking, et al. 1984)

Table 1a. Acute Toxicity of Cyanide to Freshwater Animals.

Freshwater Species	Method	LC50 or EC50 (µg CN/L)	SMAV (µg CN/L)	Analytical Method Classification	Reference
Oncorhynchus mykiss (rainbow trout)	FT, M	53		2	(McGeachy and Leduc, 1988)
Oncorhynchus mykiss (rainbow trout)	FT, M	53		2	(McGeachy and Leduc, 1988)
Oncorhynchus mykiss (rainbow trout)	FT, M	42		2	(McGeachy and Leduc, 1988)
Oncorhynchus mykiss (rainbow trout)	S, U	97*		6	(Skibba, 1981)
Oncorhynchus mykiss (rainbow trout)	FT, M	57		2	(Smith et al., 1979)
Perca flavescens (yellow perch)	FT, M	281*	93	2	(Smith et al., 1978)
Perca flavescens (yellow perch)	FT, M	288*		2	(Smith et al., 1978)
Perca flavescens (yellow perch)	FT, M	330*		2	(Smith et al., 1978)
Perca flavescens (yellow perch)	FT, M	88.9		2	(Smith et al., 1978)
Perca flavescens (yellow perch)	FT, M	93		2	(Smith et al., 1978)
Perca flavescens (yellow perch)	FT, M	74.7		2	(Smith et al., 1978)
Perca flavescens (yellow perch)	FT, M	94.7		2	(Smith et al., 1978)
Perca flavescens (yellow perch)	FT, M	101		2	(Smith et al., 1978)
Perca flavescens (yellow perch)	FT, M	107		2	(Smith et al., 1978)
Physa heterostropha (snail)	S, U	432	432	6	(Cairns and Scheier, 1958; Patrick et al., 1968)
Pimephales promelas (fathead minnow)	FT, M	120	125	2	(Broderius et al., 1977)
Pimephales promelas (fathead minnow)	FT, M	113		2	(Broderius et al., 1977)
Pimephales promelas (fathead minnow)	FT, M	128		2	(Broderius et al., 1977)
Pimephales promelas (fathead minnow)	FT, M	128		2	(Broderius et al., 1977)
Pimephales promelas (fathead minnow)	S, U	230*		6	(Doudoroff, 1956)
Pimephales promelas (fathead minnow)	S, U	170*		6	(Ewell et al. 1986)
Pimephales promelas (fathead minnow)	S, M	350*		-	(Henderson et al. 1961)
Pimephales promelas (fathead minnow)	S, M	230*		-	(Henderson et al., 1961)
Pimephales promelas (fathead minnow)	FT, M	120		2	(Smith et al., 1978)
Pimephales promelas (fathead minnow)	FT, M	98.7		2	(Smith et al., 1978)
Pimephales promelas (fathead minnow)	FT, M	81.8		2	(Smith et al., 1978)
Pimephales promelas (fathead minnow)	FT, M	110		2	(Smith et al., 1978)
Pimephales promelas (fathead minnow)	FT, M	116		2	(Smith et al., 1978)
Pimephales promelas (fathead minnow)	FT, M	119		2	(Smith et al., 1978)
Pimephales promelas (fathead minnow)	FT, M	126		2	(Smith et al., 1978)
Pimephales promelas (fathead minnow)	FT, M	81.5		2	(Smith et al., 1978)
Pimephales promelas (fathead minnow)	FT, M	124		2	(Smith et al., 1978)
Pimephales promelas (fathead minnow)	FT, M	137		2	(Smith et al., 1978)
Pimephales promelas (fathead minnow)	FT, M	131		2	(Smith et al., 1978)

Table 1a. Acute Toxicity of Cyanide to Freshwater Animals.

Freshwater Species	Method	LC50 or EC50 (µg CN/L)	SMAV (µg CN/L)	Analytical Method Classification	Reference
Pimephales promelas (fathead minnow)	FT, M	105		2	(Smith et al., 1978)
Pimephales promelas (fathead minnow)	FT, M	119		2	(Smith et al., 1978)
Pimephales promelas (fathead minnow)	FT, M	131		2	(Smith et al., 1978)
Pimephales promelas (fathead minnow)	FT, M	122		2	(Smith et al., 1978)
Pimephales promelas (fathead minnow)	FT, M	161		2	(Smith et al., 1978)
Pimephales promelas (fathead minnow)	FT, M	188		2	(Smith et al., 1978)
Pimephales promelas (fathead minnow)	FT, M	175		2	(Smith et al., 1978)
Pimephales promelas (fathead minnow)	FT, M	163		2	(Smith et al., 1978)
Pimephales promelas (fathead minnow)	FT, M	169		2	(Smith et al., 1978)
Poecilia reticulata (guppy)	FT, M	147	147	6**	(Anderson and Weber, 1975)
Pomoxis nigromaculatus (black crappie)	FT, M	102	102	2	(Smith et al., 1978)
Pteronarcys dorsata (stonefly)	FT, M	426	426	6	(Call and Brooke, 1982)
Rana pipiens (northern leopard frog)	FT, M	192	281	1	(ENSR, 2005b)
Rana berlandieri (Rio Grande leopard frog)	FT, M	410.1		1	(ENSR, 2005a)
Ranatra sp. (water stick insect)	S, U	228	230	6	(Sarkar, 1990)
Ranatra sp. (water stick insect)	S, U	229		6	(Sarkar, 1990)
Ranatra sp. (water stick insect)	S, U	231		6	(Sarkar, 1990)
Salmo salar (Atlantic salmon)	R, M	90	90	1	(Tryland and Grande, 1983)
Salvelinus fontinalis (brook trout)	FT, M	156	86	2	(Cardwell et al., 1976)
Salvelinus fontinalis (brook trout)	FT, M	105*		2	(Smith et al., 1978)
Salvelinus fontinalis (brook trout)	FT, M	342*		2	(Smith et al., 1978)
Salvelinus fontinalis (brook trout)	FT, M	507*		2	(Smith et al., 1978)
Salvelinus fontinalis (brook trout)	FT, M	252*		2	(Smith et al., 1978)
Salvelinus fontinalis (brook trout)	FT, M	84		2	(Smith et al., 1978)
Salvelinus fontinalis (brook trout)	FT, M	54.4		2	(Smith et al., 1978)
Salvelinus fontinalis (brook trout)	FT, M	86.5		2	(Smith et al., 1978)
Salvelinus fontinalis (brook trout)	FT, M	104		2	(Smith et al., 1978)
Salvelinus fontinalis (brook trout)	FT, M	90.3		2	(Smith et al., 1978)
Salvelinus fontinalis (brook trout)	FT, M	73.5		2	(Smith et al., 1978)
Salvelinus fontinalis (brook trout)	FT, M	83		2	(Smith et al., 1978)
Salvelinus fontinalis (brook trout)	FT, M	75		2	(Smith et al., 1978)
Salvelinus fontinalis (brook trout)	FT, M	86.4		2	(Smith et al, 1978)
Salvelinus fontinalis (brook trout)	FT, M	91.9		2	(Smith et al.,1978)

Table 1a. Acute Toxicity of Cyanide to Freshwater Animals.

Freshwater Species	Method	LC50 or EC50 (µg CN/L)	SMAV (µg CN/L)	Analytical Method Classification	Reference
Salvelinus fontinalis (brook trout)		FT, M		2	(Smith et al., 1978)
Salvelinus fontinalis (brook trout)	FT, M	96.7		2	(Smith et al., 1978)
Salvelinus fontinalis (brook trout)	FT, M	112		2	(Smith et al., 1978)
Salvelinus fontinalis (brook trout)	FT, M	52		2	(Smith et al., 1978)
Salvelinus fontinalis (brook trout)	FT, M	60.2		2	(Smith et al., 1978)
Salvelinus fontinalis (brook trout)	FT, M	66.8		2	(Smith et al., 1978)
Salvelinus fontinalis (brook trout)	FT, M	71.4		2	(Smith et al., 1978)
Salvelinus fontinalis (brook trout)	FT, M	97		2	(Smith et al., 1978)
Salvelinus fontinalis (brook trout)	FT, M	143		2	(Smith et al., 1978)
Tanytarsus dissimilis (midge)	S, M	2,490	2,490	2	(Call et al., 1983)
Xenopus laevis (frog)	FT, M	254.3	254.3	1	(ENSR, 2005c)

*Not used in calculation of SMAV

**Cyanide reported as measured, but insufficient information provided on the analytical method.

S = static, R = renewal, FT = flow-through; U = unmeasured, M = measured

Table 1b. Acute Toxicity of Cyanide to Saltwater Animals.

Saltwater Species	Method	LC50 or EC50 (µg CN/L)	SMAV (µg CN/L)	Analytical Method Classification	Reference
Acartia clausi (copepod)	S, U	17	17	6	(Gentile, 1980; Lussier and Cardin, 1985; Hansen and Burgess, 1987)
Americamysis bahia (mysid)	FT, M	113	113	4	(Lussier et al., 1985)
Americamysis bahia (mysid)	S, U	93*		6	(Lussier and Cardin, 1985)
Americamysis bigelowi (mysid)	S, U	124	124	6	(Lussier and Cardin, 1985)
Ampelisca abdita (amphipod)	S, U	1,220	996	6	(Scott et al., 1983)
Ampelisca abdita (amphipod)	S, U	1,150		6	(Scott et al., 1983)
Ampelisca abdita (amphipod)	S, U	704		6	(Scott et al., 1983)
Cancer gracilis (slender crab)	R, M	153	144	3	(Brix et al., 2000)
Cancer gracilis (slender crab)	R, M	135		3	(Brix et al., 2000)
Cancer irroratus (rock crab)	R, M	4.2	22	2**	(Johns and Gentile, 1981)
Cancer irroratus (rock crab)	R, M	5.7		2**	(Johns and Gentile, 1981)
Cancer irroratus (rock crab)	FT, M	70.9		1	(Northwestern Aquatic Sciences, 2005c)
Cancer irroratus (rock crab)	FT, M	44.2		1	(Northwestern Aquatic Sciences, 2005b)
Cancer irroratus (rock crab)	FT, M	70.4		1	(Northwestern Aquatic Sciences, 2005a)
Cancer magister (Dungeness crab)	R, M	51	68	3	(Brix, et al. 2000)
Cancer magister (Dungeness crab)	R, M	92		3	(Brix, et al. 2000)
Cancer oregonensis (Oregon cancer crab)	R, M	111	131	3	(Brix, et al. 2000)
Cancer oregonensis (Oregon cancer crab)	R, M	154		3	(Brix, et al. 2000)
Cancer productus (red crab)	R, M	219	153	3	(Brix, et al. 2000)
Cancer productus (red crab)	R, M	107		3	(Brix, et al. 2000)
Crepidula fornicata (Atlantic slippershell)	S, U	>10,000	>10,000	6	(Gardner and Nelson, 1981)
Cyprinodon variegatus (sheepshead minnow)	FT, M	300	300	2**	(Schimmel, 1981)
Dinophilus gyrociliatus (archiannelid)	R, M	3,152	3,560	6	(Carr et al., 1986)
Dinophilus gyrociliatus (archiannelid)	R, M	4,021		6	(Carr et al., 1986)
Menidia menidia (Atlantic silverside)	FT, M	59	59	2**	(Gardner and Berry, 1981)
Pseudopleuronectes americanus (winter flounder)	S, U	372	372	6	(Cardin, 1980)

*Not used in calculation of SMAV

** = Assumed free CN tested, based on species reported and method employed in other studies conducted at U.S. EPA laboratories.

S = static, R = renewal, FT = flow-through

U = unmeasured, M = measured

Table 2a. Chronic Toxicity of Cyanide to Freshwater Animals.

Freshwater Species	Test	Limits (µg CN/L)	Chronic Value (µg CN/L)	EC20 (µg CN/L)	Analytical Method Classification	Reference
Asellus communis (isopod)	LC	29-40	34.06	32	2	(Oseid and Smith, 1979)
Gammarus pseudolimnaeus (amphipod)	LC	16-21	18.33	14.33	2	(Oseid and Smith, 1979)
Lepomis macrochirus (bluegill)	ELS	9.3-19.8	13.57	5.595	2	(Kimball et al. 1978)
Pimephales promelas (fathead minnow)	LC	13.3-20.2	16.39	6.038	2	(Lind et al., 1977)
Salvelinus fontinalis (brook trout)	LC	5.6-11.0	7.849	7.705	2	(Koenst et al., 1977)

LC = life cycle or partial life cycle

ELS = early life stage

Table 2b. Chronic toxicity of cyanide to saltwater animals.

Saltwater Species	Test	Limits (µg CN/L)	Chronic Value (µg CN/L)	EC20	Analytical Method Classification	Reference
Americamysis bahia (mysid)	LC	43-113	69.71	43*	2	(Lussier et al., 1985)
Cyprinodon variegatus (sheepshead minnow)	ELS	29-45	36.12	29*	2**	(Schimmel, 1981)

LC = life cycle or partial life cycle

ELS = early life stage

* = Assumed to be similar to the NOEC (see text).

** = Assumed free CN tested, based on species reported and method employed in other studies conducted at U.S. EPA laboratories.

Table 2c. Cyanide Acute-chronic Ratios.

Species	Reference	Acute Value (µg CN/L)	Chronic Value* (µg CN/L)	ACR	1984 ACR
FRESHWATER SPECIES					
Asellus communis (isopod)	(Oseid and Smith, 1979)	2,326	31.65	73.49	68.29
Gammarus pseudolimnaeus (amphipod)	(Oseid and Smith, 1979)	167	14.31	11.67	9.111
Salvelinus fontinalis (brook trout)	(Koenst et al., 1977)	85.8	7.705	11.14	10.59
Pimephales promelas (fathead minnow)	(Lind et al., 1977)	125.1	6.038	20.72	7.633
Lepomis macrochirus (bluegill)	(Kimball et al., 1978)	99.28	8.941	11.10	7.316
SALTWATER SPECIES					
Americamysis bahia (mysid)	(Lussier et al., 1985)	113	43	2.627	1.621
Cyprinodon variegates (sheepshead minnow)	(Schimmel, 1981)	300	29	10.34	8.306
		Geometric mean of freshwater ACRs** =		13.15	
		Geometric mean of saltwater ACRs =		5.212	
		Geometric mean of freshwater and saltwater ACRs** =		9.659	

* Chronic value is expressed as an EC20 when possible, but not in all cases (see text).

** Excluding the *Asellus communis* ACR

Table 3a. Ranked Freshwater Genus Mean Acute Values with Species Mean Acute-Chronic Ratios.

Rank	GMAV (µg CN/L)	Freshwater Species	SMAV (µg CN/L)	Species Mean ACR
28	>50,000	*Helisoma trivolvis* (snail)	>50,000	-
27	11,000	*Lumbriculus variegatus* (oligochaete)	11,000	-
26	2,490	*Tanytarsus dissimilis* (midge)	2,490	-
25	2,100	*Dugesia tigrina* (planarian)	2,100	-
24	1,989	*Asellus communis* (isopod)	2,326	73.49
		Asellus intermedius (isopod)	1,700	-
23	432.0	*Physa heterostropha* (snail)	432.0	-
22	426.0	*Pteronarcys dorsata* (stonefly)	426.0	-
21	387.7	*Gammarus fasciatus* (amphipod)	900.0	-
		Gammarus pseudolimnaeus (amphipod)	167.0	11.65
20	318.0	*Carassius auratus* (goldfish)	318.0	-
19	280.6	*Rana pipiens* (northern leopard frog)	192.0	-
		Rana berlandieri (Rio Grande leopard frog)	410.1	-
18	273.3	*Nepa sp.* (water scorpion)	273.3	-
17	254.3	*Xenopus laevis* (frog)	254.3	-
16	251.3	*Dytiscus sp.* (aquatic beetle)	251.3	-
15	249.8	*Corixa sp.* (water boatman)	249.8	-
14	229.6	*Ranatra sp.* (water stick insect)	229.6	-
13	147.0	*Poecilia reticulata* (guppy)	147.0	-
12	132.6	*Diaptomus sp.* (copepod)	132.6	-
11	129.9	*Cyclops viridis* (copepod)	129.9	-
10	125.1	*Pimephales promelas* (fathead minnow)	125.1	20.72
9	114.9	*Daphnia spp.* (cladoceran)	132.3	-
		Daphnia magna (cladoceran)	120.0	-
		Daphnia pulex (cladoceran)	95.55	-

Table 3a. Ranked Freshwater Genus Mean Acute Values with Species Mean Acute-Chronic Ratios.

Rank	GMAV (µg CN/L)	Freshwater Species	SMAV (µg CN/L)	Species Mean ACR
8	102.0	*Micropterus salmoides* (largemouth bass)	102.0	-
7	102.0	*Pomoxis nigromaculatus* (black crappie)	102.0	-
6	99.28	*Lepomis macrochirus* (bluegill)	99.28	17.74
5	98.80	*Gasterosteus aculeatus* (three-spine stickleback)	98.80	-
4	92.64	*Perca flavescens* (yellow perch)	92.64	-
3	90.00	*Salmo salar* (Atlantic salmon)	90.00	-
2	85.80	*Salvelinus fontinalis* (brook trout)	85.80	11.14
1	46.53	*Oncorhynchus mykiss* (rainbow trout)	46.53	-

Table 3b. Ranked Saltwater Genus Mean Acute Values with Species Mean Acute-Chronic Ratios.

Rank	GMAV (µg CN/L)	Saltwater Species	SMAV (µg CN/L)	Species Mean ACR
9	>10,000	*Crepidula fornicata* (Atlantic slippershell)	>10,000	-
8	3,560	*Dinophilus gyrociliatus* (archiannelid)	3,560	-
7	996	*Ampelisca abdita* (amphipod)	995.9	-
6	372	*Pseudopleuronectes americanus* (winter flounder)	372.0	-
5	300	*Cyprinodon variegatus* (sheepshead minnow)	300.0	10.34
4	118.4	*Americamysis bigelowi* (mysid)	124.0	2.627
		Americamysis bahia (mysid)	113.0	-
3	84.69	*Cancer productus* (red crab)	153.1	-
		Cancer gracilis (slender crab)	143.7	-
		Cancer oregonensis (Oregon cancer crab)	130.7	-
		Cancer irroratus (rock crab)	22.11	-
		Cancer magister (Dungeness crab)	68.50	-
2	59.00	*Menidia menidia* (Atlantic silverside)	59.00	-
1	17.00	*Acartia clausi* (copepod)	17.00	-

Table 3c. Freshwater and Saltwater Cyanide Criteria Values.

Freshwater

CMC:

Considering GMAVs for Coldwater and Warmwater Species:

Final Acute Value = 58.95 µg CN/L (calculated from Genus Mean Acute Values)

Final Acute Value = 46.53 µg CN/L (lowered to protect rainbow trout; see text)

Criterion Maximum Concentration = 46.53 µg CN/L / 2 = 23.27 µg CN/L

Final Acute-Chronic Ratio = 9.659 (see text)

Considering GMAVs for Warmwater Species (Excluding Salmonid GMAVs):

Final Acute Value = 94.50 µg CN/L (calculated from Genus Mean Acute Values)

Criterion Maximum Concentration = 94.50 µg CN/L / 2 = 47.25 µg CN/L

Final Acute-Chronic Ratio = 9.659 (see text)

CCC:

Considering GMAVs for Coldwater and Warmwater Species:

Final Chronic Value = (46.53 µg CN/L) / 9.659 = 4.817 µg CN/L

Considering GMAVs for Warmwater Species (Excluding Salmonid GMAVs):

Final Chronic Value = (94.50 µg CN/L) / 9.659 = 9.784 µg CN/L

Saltwater

CMC:

Final Acute Value = 11.00 µg CN/L (calculated from Genus Mean Acute Values)

Criterion Maximum Concentration = 11.00 µg CN/L / 2 = 5.500 µg CN/L

Final Acute-Chronic Ratio = 9.659 (see text)

CCC:

Final Chronic Value = (11.00 µg CN/L) / 9.659 = 1.139 µg CN/L

Table 4a. Toxicity of Cyanide to Freshwater Plants.

Freshwater Species	Effect	Result (µg CN/L)	Reference
Chlamydomonas sp. (green algae)	No effect on growth rate	10-100	(Cairns et al., 1978)
Lemna gibba (duckweed)	Decreased potassium uptake	26,000	(Kondo and Tsudzuki, 1980)
Microcystis aeruginosa (blue-green alga)	LOEC for population growth	75	(Bringmann, 1975; Bringmann and Kuhn 1976, 1978a, 1978b)
Microcystis aeruginosa (blue-green alga)	90% mortality (LC90)	8,000	(Fitzgerald et al., 1952)
Myriophyllum spicatum (Eurasian watermilfoil)	EC50 for root weight	22,400	(Stanley, 1974)
Navicula seminulum (diatom)	EC50 for population growth	277-491	(Academy of Natural Sciences, 1960)
Scenedesmus quadricauda (green alga)	LOEC for population growth	30	(Bringmann and Kuhn, 1977a; Bringmann and Kuhn, 1978a, 1978b, 1980a)

LOEC = lowest observed effect concentration

LC90 = Lethal concentration to 90% of test population

EC50 = Median effect concentration

Table 4b. Toxicity of Cyanide to Saltwater Plants.

Saltwater Species	Effect	Result (µg CN/L)	Reference
Champia parvula (red alga)	Reduced tetrasporophyte growth	16	(Steele and Thursby, 1983)
Champia parvula (red alga)	Reduced tetrasporophyte production	25	(Steele and Thursby, 1983)
Champia parvula (red alga)	Reduced female growth	11	(Steele and Thursby, 1983)
Champia parvula (red alga)	Stopped sexual reproduction	11	(Steele and Thursby, 1983)
Chlorella sp. (green alga)	Enzyme inhibition	30,000	(Nelson and Tolbert, 1970)
Nitzschia closterium (diatom)	EC50 for population growth	30	(Pablo et al., 1997b)
Nitzschia closterium (diatom)	LOEC for population growth	5.3	(Pablo et al., 1997b)
Prototheca zopti (green alga)	Respiration inhibition	3,000	(Webster and Hackett, 1965)

LOEC = lowest observed effect concentration

EC50 = Median effect concentration

Table 5a. Other Data on Effects of Cyanide on Freshwater Organisms.

Freshwater Species	Exposure Duration	Effect	Result (µg CN/L)	Reference
	Studies from Table 5 of the 1984 AWQC Document:			
Scenedesmus quadricauda (green alga)	96 hrs	Growth	160*	(Bringmann & Kuhn, 1959a, 1959b)
Escherichia coil (bacteria)	-	Incipient inhibition	400-800	(Bringmann & Kuhn, 1959a)
Pseudomonas putida (bacteria)	16 hrs	Incipient inhibition	1	(Bringmann & Kuhn, 1976, Bringmann & Kuhn, 1977a, 1979, Bringmann & Kuhn, 1980a)
Entosiphon sulcatum (protozoan)	72 hrs	Incipient inhibition	1,800	(Bringmann & Kuhn, 1978a, Bringmann & Kuhn, 1979, Bringmann & Kuhn, 1980b, 1981)
Microregma heterostoma (protozoan)	28 hrs	Incipient inhibition	40	(Bringmann & Kuhn, 1959b)
Chilomonas paramecium (protozoan)	48 hrs	Incipient inhibition	1,200	(Bringmann & Kuhn, 1980b, 1981)
Uronema parduezi (protozoan)	20 hrs	Incipient inhibition	270	(Bringmann & Kuhn, 1980a, 1981)
Philodina acuticornis (rotifer)	48 hrs	LC50	20,000-145,000	(Cairns et al., 1978)
Aeolosoma headleyi (worm)	48 hrs	LC50 (5°C)	10,000	(Cairns et al., 1978)
Aeolosoma headleyi (worm)	48 hrs	LC50 (10°C)	9,000	(Cairns et al., 1978)
Aeolosoma headleyi (worm)	48 hrs	LC50 (15°C)	120,000	(Cairns et al., 1978)
Aeolosoma headleyi (worm)	48 hrs	LC50 (20°C)	160,000	(Cairns et al., 1978)
Aeolosoma headleyi (worm)	48 hrs	LC50 (25°C)	160,000	(Cairns et al., 1978)
Goniobasis livescens (snail)	48 hrs	LC50	760,000	(Cairns, 1976)
Nitocris sp. (snail)	48 hrs	LC50 (5°C)	13,600	(Cairns et al., 1978)
Nitocris sp. (snail)	48 hrs	LC50 (10°C)	12,800	(Cairns et al., 1978)
Nitocris sp. (snail)	48 hrs	LC50 (15°C)	10,000	(Cairns et al., 1978)
Nitocris sp. (snail)	48 hrs	LC50 (20°C)	8,000	(Cairns et al., 1978)
Nitocris sp. (snail)	48 hrs	LC50 (25°C)	7,000	(Cairns et al., 1978)
Lymnaea emarginata (snail)	48 hrs	LC50	3,300	(Cairns, 1976)
Lymnaea sp. (snail, embryo)	96 hrs	LC50	52,000	(Dowden and Bennett, 1965)
Physa heterostropha (snail)	96 hrs	LC50 (periodic low DO)	190	(Cairns and Scheier, 1958)
Physa integra (snail)	48 hrs	LC50	1,350	(Cairns, 1976)
Daphnia magna (cladoceran)	48 hrs	EC50	800*	(Bringmann & Kuhn, 1959a, 1959b)

Table 5a. Other Data on Effects of Cyanide on Freshwater Organisms.

Freshwater Species	Exposure Duration	Effect	Result (µg CN/L)	Reference
Daphnia magna (cladoceran)	24 hrs	LC50	530	(Bringmann and Kuhn, 1977b)
Daphnia pulex (cladoceran)	48 hrs	LC50 (5°C)	330	(Cairns et al., 1978)
Daphnia pulex (cladoceran)	48 hrs	LC50 (10°C)	330	(Cairns et al., 1978)
Daphnia pulex (cladoceran)	48 hrs	LC50 (15°C)	180	(Cairns et al., 1978)
Daphnia pulex (cladoceran)	48 hrs	LC50 (25°C)	1	(Cairns et al., 1978)
Gammarus pseudolimnaeus (amphipod)	98 days	Competition with *Asellus* affects HCN toxicity	9	(Oseid and Smith, 1979)
Stenonema rubrum (mayfly)	48 hrs	LC50	500	(Roback, 1965)
Hydropsyche sp. (caddisfly larva)	48 hrs	LC50	2,000	(Roback, 1965)
Tanytarsus dissimilis (midge)	48 hrs	EC50	<880	(Call, 1979)
Carassius auratus (goldfish, juvenile)	336 hrs	LC50	261	(Cardwell et al., 1976)
Carassius auratus (goldfish, juvenile)	24 hrs	LC50 (5°C)	3,250	(Cairns et al., 1978)
Carassius auratus (goldfish, juvenile)	24 hrs	LC50 (15°C)	440	(Cairns et al., 1978)
Carassius auratus (goldfish, juvenile)	24 hrs	LC50 (30°C)	280	(Cairns et al., 1978)
Gambusia affinis (mosquitofish)	96 hrs	LC50 (high turbidity)	640	(Wallen et al,. 1957)
Gasterosteus aculeatus (threespine stickleback)	90 min	Depressed respiratory rate to 32% of normal	1,040	(Jones, 1947)
Gasterosteus aculeatus (threespine stickleback, adult)	824 min	Median survival time	134	(Broderius, 1973)
Gasterosteus aculeatus (threespine stickleback, adult)	642 min	Median survival time	170	(Broderius, 1973)
Gasterosteus aculeatus (threespine stickleback, adult)	412 min	Median survival time	237	(Broderius, 1973)
Ichtalurus punctatus (channel catfish, juvenile)	26 hrs	LC50	161	(Cardwell et al.,1976)
Ichtalurus punctatus (channel catfish, juvenile)	24 hrs	LC50 (5°C)	200	(Cairn et al., 1978)
Ichtalurus punctatus (channel catfish, juvenile)	24 hrs	LC50 (15°C)	310	(Cairns et al., 1978)
Ichtalurus punctatus (channel catfish, juvenile)	24 hrs	LC50 (30°C)	230	(Cairns,et al., 1978)
Jordanella floridae (flagfish)	10 days	Reduced fecundity and hatching	63	(Cheng and Ruby, 1981)
Lepomis machrochirus (bluegill, adult)	48 hrs	LC50	160	(Cairns, 1965)
Lepomis machrochirus (bluegill, adult)	289 days	Survival reduced	67.8	(Kimball et al., 1978)
Lepomis machrochirus (bluegill, adult)	289 days	No reproduction	5.4	(Kimball et al., 1978)

Table 5a. Other Data on Effects of Cyanide on Freshwater Organisms.

Freshwater Species	Exposure Duration	Effect	Result (µg CN/L)	Reference
Lepomis machrochirus (bluegill, juvenile)	202 min	Median survival time	198	(Broderius, 1973)
Lepomis machrochirus (bluegill, juvenile)	260 min	Median survival time	194	(Broderius, 1973)
Lepomis machrochirus (bluegill, juvenile)	351 min	Median survival time	165	(Broderius, 1973)
Lepomis machrochirus (bluegill, juvenile)	258 min	Median survival time	165	(Broderius, 1973)
Lepomis machrochirus (bluegill, juvenile)	352 min	Median survival time	144	(Broderius, 1973)
Lepomis machrochirus (bluegill, juvenile)	655 min	Median survival time	127	(Broderius, 1973)
Lepomis machrochirus (bluegill, juvenile)	48 hrs	LC50	134	(Cardwell et al., 1976)
Lepomis machrochirus (bluegill, juvenile)	48 hrs	LC50	280	(Turnbull, 1954)
Lepomis machrochirus (bluegill, juvenile)	50 min	Median resistance time	960	(Doudoroff et al., 1966)
Lepomis machrochirus (bluegill, juvenile)	91 min	Median resistance time	720	(Doudoroff et al., 1966)
Lepomis machrochirus (bluegill, juvenile)	129 min	Median resistance time	540	(Doudoroff et al., 1966)
Lepomis machrochirus (bluegill, juvenile)	700 min	Median resistance time	170	(Doudoroff et al., 1966)
Lepomis machrochirus (bluegill, juvenile)	72 hrs	LC50	154	(Doudoroff et al., 1966)
Lepomis machrochirus (bluegill, juvenile)	24 hrs	LC50 (5°C)	240	(Cairns et al., 1978)
Lepomis machrochirus (bluegill, juvenile)	24 hrs	LC50 (15°C)	160	(Cairns et al., 1978)
Lepomis machrochirus (bluegill, juvenile)	24 hrs	LC50 (30°C)	190	(Cairns et al., 1978)
Lepomis machrochirus (bluegill, juvenile)	96 hrs	LC50 (periodic low DO)	48	(Cairns and Scheier, 1958)
Micropterus dolomieui (smallmouth bass, juvenile)	7.8 min	Geometric mean time to death	1,900	(Burdick, 1958)
Micropterus dolomieui (smallmouth bass, juvenile)	12.4 min	Geometric mean time to death	1,430	(Burdick, 1958)
Micropterus dolomieui (smallmouth bass, juvenile)	15.4 min	Geometric mean time to death	978	(Burdick, 1958)
Micropterus dolomieui (smallmouth bass, juvenile)	30.6 min	Geometric mean time to death	755	(Burdick, 1958)
Micropterus dolomieui (smallmouth bass, juvenile)	42.8 min	Geometric mean time to death	478	(Burdick, 1958)
Micropterus dolomieui (smallmouth bass, juvenile)	80.5 min	Geometric mean time to death	338	(Burdick, 1958)
Micropterus dolomieui (smallmouth bass, juvenile)	122 min	Geometric mean time to death	243	(Burdick, 1958)
Micropterus dolomieui (smallmouth bass, juvenile)	290 min	Geometric mean time to death	175	(Burdick, 1958)
Micropterus salmoides (largemouth bass, juvenile)	2 days	Significant increases in opercular rate	40	(Morgan and Kuhn, 1974)
Micropterus salmoides (largemouth bass, juvenile)	24 hrs	Affected opercular rhythm	10	(Morgan, 1979)
Notemigonus crysoleucas (golden shiner, juvenile)	24 hrs	LC50 (5°C)	540	(Cairns et al., 1978)
Notemigonus crysoleucas (golden shiner, juvenile)	24 hrs	LC50 (15°C)	310	(Cairns et al., 1978)

Table 5a. Other Data on Effects of Cyanide on Freshwater Organisms.

Freshwater Species	Exposure Duration	Effect	Result (µg CN/L)	Reference
Notemigonus crysoleucas (golden shiner, juvenile)	24 hrs	LC50 (30°C)	300	(Cairns et al., 1978)
Oncorhynchus kisutch (coho salmon)	2 hrs	Swimming speed reduced	10	(Broderius, 1970)
Oncorhynchus kisutch (coho salmon, juvenile)	36 days	Reduction in growth	77	(Leduc, 1966)
Oncorhynchus mykiss (rainbow trout)	48 hrs	LC50	68	(Brown, 1968)
Oncorhynchus mykiss (rainbow trout, adult)	2 min	Mean survival time	2,000	(Herbert and Merkens, 1952)
Oncorhynchus mykiss (rainbow trout, adult)	8 min	Mean survival time	300	(Herbert and Merkens, 1952)
Oncorhynchus mykiss (rainbow trout, adult)	12 min	Mean survival time	250	(Herbert and Merkens, 1952)
Oncorhynchus mykiss (rainbow trout, adult)	12 min	Mean survival time	200	(Herbert and Merkens, 1952)
Oncorhynchus mykiss (rainbow trout, adult)	24 min	Mean survival time	180	(Herbert and Merkens, 1952)
Oncorhynchus mykiss (rainbow trout, adult)	72 min	Mean survival time	160	(Herbert and Merkens, 1952)
Oncorhynchus mykiss (rainbow trout, adult)	90 min	Mean survival time	140	(Herbert and Merkens, 1952)
Oncorhynchus mykiss (rainbow trout, adult)	2,525 min	Mean survival time	100	(Herbert and Merkens, 1952)
Oncorhynchus mykiss (rainbow trout, adult)	1,617 min	Mean survival time	90	(Herbert and Merkens, 1952)
Oncorhynchus mykiss (rainbow trout, adult)	3,600 min	Mean survival time	80	(Herbert and Merkens, 1952)
Oncorhynchus mykiss (rainbow trout, adult)	4,441 min	Mean survival time	70	(Herbert and Merkens, 1952)
Oncorhynchus mykiss (rainbow trout, juvenile)	250 min	Approximate median survival time	200	(Department of Scientific and Industrial Research, 1956)
Oncorhynchus mykiss (rainbow trout, juvenile)	18 days	Weight gain reduced	9.6	(Dixon and Leduc, 1981)
Oncorhynchus mykiss (rainbow trout, juvenile)	4 days	Increased respiration rate	9.6	(Dixon and Leduc, 1981)
Oncorhynchus mykiss (rainbow trout, juvenile)	18 days	Liver damage (necrobiosis)	9.6	(Dixon and Leduc, 1981)
Oncorhynchus mykiss (rainbow trout, juvenile)	18 days	Reduction in fat content	19	(Dixon and Leduc, 1981)
Oncorhynchus mykiss (rainbow trout, juvenile)	18 days	Higher relative body water content	9.6	(Dixon and Leduc, 1981)
Oncorhynchus mykiss (rainbow trout, juvenile)	28 days	Altered blood chloride and osmolarity	9.6	(Leduc and Chan, 1975)
Oncorhynchus mykiss (rainbow trout, juvenile)	18 days	Production of dividing spermatogonia reduced by 13%	9.6	(Ruby, 1979)
Oncorhynchus mykiss (rainbow trout, juvenile)	18 days	Production of dividing spermatogonia reduced by 50%	29	(Ruby, 1979)
Oncorhynchus mykiss (rainbow trout, juvenile)	24 hrs	LC50 (5°C)	90	(Cairns et al., 1978)
Oncorhynchus mykiss (rainbow trout, juvenile)	24 hrs	LC50 (12°C)	98	(Cairns et al., 1978)
Oncorhynchus mykiss (rainbow trout, juvenile)	24 hrs	LC50 (18°C)	92	(Cairns et al., 1978)

Table 5a. Other Data on Effects of Cyanide on Freshwater Organisms.

Freshwater Species	Exposure Duration	Effect	Result (µg CN/L)	Reference
Oncorhynchus mykiss (rainbow trout, juvenile)	21 days	No effect on dry weight gain	33	(Dixon and Sprague, 1981)
Oncorhynchus mykiss (rainbow trout, juvenile)	21 days	Kidney damage	33	(Dixon and Sprague, 1981)
Oncorhynchus mykiss (rainbow trout, juvenile)	144 hrs	LC50	93	(Dixon and Sprague, 1981)
Oncorhynchus mykiss (rainbow trout, juvenile)	20 days	Reduction in swimming ability (6-18°C)	4.8-43	(Kovacs, 1979; Kovacs and Leduc 1982b)
Oncorhynchus mykiss (rainbow trout, juvenile)	20 days	Threshold concentration (6-18°C) for reduction of relative: wet weight gain dry weight gain fat gain	 9.6-29 <4.8-29 <4.8-24	(Kovacs 1979; Kovacs and Leduc, 1982b)
Oncorhynchus mykiss (rainbow trout, juvenile)	20 days	Increase in relative water content (6-18°C)	4.8-43	(Kovacs, 1979; Kovacs and Leduc, 1982b)
Oncorhynchus mykiss (rainbow trout, juvenile)	20 days	No effect on wet or dry weight relative growth rate or fat weight change for 8 g fish forced to swim at 12 cm/sec and 10°C	9.6	(McCracken and Leduc, 1980)
Oncorhynchus mykiss (rainbow trout, juvenile)	20 days	Decreased wet weight gain for 27 g fish forced to swim at 12 cm/sec and 10°C	9.6	(McCracken and Leduc, 1980)
Oncorhynchus mykiss (rainbow trout, yearling)	21 days	65% reduction in weight gain	19	(Speyer, 1975)
Oncorhynchus mykiss (rainbow trout, yearling)	21 days	75% reduction in swimming ability	19	(Speyer, 1975)
Oncorhynchus mykiss (rainbow trout, yearling)	21 days	Higher relative body water content	19	(Speyer, 1975)
Oncorhynchus mykiss (rainbow trout, yearling)	20 days	Abnormal oocyte development	9.6	(Lesniak, 1977; Lesniak and Ruby, 1982)
Oncorhynchus mykiss (rainbow trout, yearling)	7 days	Serum calcium reduced; hepatosomatic indices declined	9.6 19	(Costa and Ruby, 1984)
Oncorhynchus mykiss (rainbow trout, yearling)	20 days	Increased food maintenance requirements, decreased wet and dry weight relative growth rate and fat weight change for 18 g fish forced to swim at 12 cm/sec and 10°C	13	(McCracken and Leduc, 1980)
Oncorhynchus tshawytscha (chinook salmon, juvenile)	64 days	27% reduction in biomass	20	(Negilski, 1973)
Pimephales promelas (fathead minnow)	48 hrs	LC50	240	(Black, 1957)
Pimephales promelas (fathead minnow, embryo)	96 hrs	LC50	347	(Smith et al., 1978)
Pimephales promelas (fathead minnow, embryo)	96 hrs	LC50	272	(Smith et al., 1978)

Table 5a. Other Data on Effects of Cyanide on Freshwater Organisms.

Freshwater Species	Exposure Duration	Effect	Result (µg CN/L)	Reference
Pimephales promelas (fathead minnow, embryo)	96 hrs	LC50	201	(Smith et al., 1978)
Pimephales promelas (fathead minnow, embryo)	96 hrs	LC50	123	(Smith et al., 1978)
Pimephales promelas (fathead minnow, embryo)	96 hrs	LC50	186	(Smith et al., 1978)
Pimephales promelas (fathead minnow, embryo)	96 hrs	LC50	200	(Smith et al., 1978)
Pimephales promelas (fathead minnow, embryo)	96 hrs	LC50	206	(Smith et al., 1978)
Pimephales promelas (fathead minnow, juvenile)	5 days	LC50	120	(Cardwell et al., 1976)
Pimephales promelas (fathead minnow, juvenile)	10 days	LC50	114	(Cardwell et al., 1976)
Pimephales promelas (fathead minnow, juvenile)	28 days	Reduced increase in length	35	(Lind et al., 1977)
Pimephales promelas (fathead minnow, juvenile)	56 days	Reduced increase in length and weight	62	(Lind et al., 1977)
Poecilia reticulata (guppy, juvenile)	120 hrs	Threshold concentration	236	(Chen and Selleck, 1969)
Rhinichthys atratulus (blacknose dace)	24 hrs	LC50	220	(Lipschuetz and Cooper, 1955)
Salmo salar (Atlantic salmon, larva)	58 days	Abnormal embryo and larval development	9.6	(Leduc, 1978)
Salmo salar (Atlantic salmon, smolt)	24 hrs	LC50 (10 mg DO/L)	70	(Alabaster, 1983)
Salmo salar (Atlantic salmon, smolt)	24 hrs	LC50 (3.5 mg DO/L)	23	(Alabaster, 1983)
Salmo trutta (brown trout, fry)	8.2 min	Death	8,030	(Karsten, 1934)
Salmo trutta (brown trout, fry)	8.9 min	Death	4,140	(Karsten, 1934)
Salmo trutta (brown trout, fry)	8.2 min	Death	2,070	(Karsten, 1934)
Salmo trutta (brown trout, fry)	140 min	Death	217	(Karsten, 1934)
Salmo trutta (brown trout, juvenile)	6.58 min	Geometric mean time to death	1,006	(Burdick, 1958)
Salmo trutta (brown trout, juvenile)	15 min	Geometric mean time to death	510	(Burdick, 1958)
Salmo trutta (brown trout, juvenile)	30.1 min	Geometric mean time to death	320	(Burdick, 1958)
Salmo trutta (brown trout, juvenile)	5 hrs	Oxygen uptake inhibited	25	(Carter, 1962)
Salvelinus fontinalis (brook trout, fry)	15.2 min	Death	8,640	(Karsten, 1934)
Salvelinus fontinalis (brook trout, fry)	10.8 min	Death	4,290	(Karsten, 1934)
Salvelinus fontinalis (brook trout, fry)	11.7 min	Death	2,130	(Karsten, 1934)
Salvelinus fontinalis (brook trout, fry)	26 min	Death	853	(Karsten, 1934)
Salvelinus fontinalis (brook trout, fry)	58 min	Death	392	(Karsten, 1934)
Salvelinus fontinalis (brook trout, fry)	210 min	Death	217	(Karsten 1934)
Salvelinus fontinalis (brook trout, fry)	130 hrs	Death	50	(Karsten, 1934)
Salvelinus fontinalis (brook trout, fry)	27 days	100% survival	20	(Karsten, 1934)

Table 5a. Other Data on Effects of Cyanide on Freshwater Organisms.

Freshwater Species	Exposure Duration	Effect	Result (µg CN/L)	Reference
Salvelinus fontinalis (brook trout, juvenile)	3.6 days	Death	80	(Neil, 1957)
Salvelinus fontinalis (brook trout, juvenile)	40 days	No death	50	(Neil, 1957)
Salvelinus fontinalis (brook trout, juvenile)	25.5 min	75% reduction in swimming endurance	10	(Neil, 1957)
Salvelinus fontinalis (brook trout, juvenile)	90 days	Reduced growth	33	(Koenst et al., 1977)
		wet weight gain	9.6-29	
		dry weight gain	<4.8-29	
		fat gain	<4.8-24	
	Studies Published After the 1984 AWQC Document:			
Spirostomum ambiguum (protozoan)	48 hrs	LC50	815	(Nalecz-Jawecki & Sawicki, 1998)
Brachionus calyciflorus (rotifer)	24 hrs	LC50	62,446	(Calleja et al., 1994)
Daphnia magna (cladoceran)	24 hrs	EC50: Immobilization	401	(Calleja et al., 1994)
Daphnia magna (cladoceran)	24 hrs	EC50: Immobilization	243	(Lilius et al., 1994)
Daphnia pulex (cladoceran)	24 hrs	EC50: Immobilization	223	(Lilius et al., 1995)
Gammarus pulex (amphipod)	11.6 to 19.4 hrs	LT50	1,199	(Abel and Garner, 1986)
Gammarus pulex (amphipod)	7 to 11.6 hrs	LT50	2,997	(Abel and Garner, 1986)
Gammarus pulex (amphipod)	4.9 to 7.8 hrs	LT50	5,993	(Abel and Garner, 1986)
Gammarus pulex (amphipod)	4.1 to 7.9 hrs	LT50	11,987	(Abel and Garner, 1986)
Gammarus pulex (amphipod)	2.5-3.4 hrs	LT50	29,967	(Abel and Garner, 1986)
Orconectes rusticus (crayfish)	96 hrs	LC100	5000	(Bills and Marking, 1988)
Gnathonemus petersi (elephantnose fish)	6 hrs	LOEC: Gill damage	20	(Alazemi et al., 1996)
Lepomis macrochirus (bluegill sunfish)	96 hrs	LC100	1000	(Bills and Marking, 1988)
Oncorhynchus mykiss (rainbow trout)	10 d	LOEC: Reproduction	0.4	(Billard and Roubaud, 1985)
Oncorhynchus mykiss (rainbow trout)	42 d	NOEC: Susceptibility to Saprolegnia parasitica	21	(Carballo et al., 1992)
Oncorhynchus mykiss (rainbow trout)	7 d	NOEC: Hepatosomatic index	>29	(Costa and Ruby, 1984)
Oncorhynchus mykiss (rainbow trout)	7 d	LOEC: Serum calcium levels reduced	>9.7	(Costa and Ruby, 1984)

Table 5a. Other Data on Effects of Cyanide on Freshwater Organisms.

Freshwater Species	Exposure Duration	Effect	Result (µg CN/L)	Reference
Oncorhynchus mykiss (rainbow trout)	7 d	NOEC: Serum phosphoprotein phosphorus	>29	(Costa and Ruby, 1984)
Oncorhynchus mykiss (rainbow trout)	96 hrs	NOEC: Liver effects	13.3	(Dixon et al., 1987)
Oncorhynchus mykiss (rainbow trout)	16 wks	NOEC: Growth	>25.5	(Lanno and Dixon, 1996)
Oncorhynchus mykiss (rainbow trout)	16 wks	NOEC: Hepatosomatic index	>25.5	(Lanno and Dixon, 1996)
Oncorhynchus mykiss (rainbow trout)	16 wks	NOEC: Splenosomatic index	>25.5	(Lanno and Dixon, 1996)
Oncorhynchus mykiss (rainbow trout)	16 wks	NOEC: Hematocrit	>25.5	(Lanno and Dixon, 1996)
Oncorhynchus mykiss (rainbow trout)	16 wks	NOEC: Plasma thyroid levels	>25.5	(Lanno and Dixon, 1996)
Oncorhynchus mykiss (rainbow trout)	21 d	LOEC: Increased plasma cortisol	<21	(Munoz et al., 1991)
Oncorhynchus mykiss (rainbow trout)	21 d	NOEC: Blood cholesterol and sodium	>21	(Munoz et al., 1991)
Oncorhynchus mykiss (rainbow trout)	20 d	LOEC: Liver cytochrome oxidase activity	<9.6	(Raymond and Leduc, 1985)
Oncorhynchus mykiss (rainbow trout)	12 d	LOEC (only conc. tested): Plasma vitellogenin	<9.6	(Ruby et al., 1986)
Oncorhynchus mykiss (rainbow trout)	12 d	LOEC (only conc. tested): Gonadosomatic index	<9.6	(Ruby et al., 1986)
Oncorhynchus mykiss (rainbow trout)	12 d	NOEC (only conc. tested): Hepatosomatic index	>9.6	(Ruby et al., 1986)
Oncorhynchus mykiss (rainbow trout)	12 d	LOEC: Number of spermatogonial cysts	<9.6	(Ruby et al., 1993)
Oncorhynchus mykiss (rainbow trout)	12 d	NOEC (only conc. tested): Gonadosomatic index	>9.6	(Ruby et al., 1993)
Oncorhynchus mykiss (rainbow trout)	12 d	LOEC (only conc. tested): Diameters of vitellogenic oocytes	<10	(Szabo et al., 1991)
Oncorhynchus mykiss (rainbow trout)	12 d	LOEC (only conc. tested): Dopamine levels	<10	(Szabo et al., 1991)
Oncorhynchus mykiss (rainbow trout)	12 d	NOEC (only conc. tested): Norepinephrine levels	>10	(Szabo et al., 1991)
Oncorhynchus mykiss (rainbow trout)	12 d	LOEC (only conc. tested): Number of spermatogonial cysts	<10	(Szabo et al., 1991)
Perca fluviatilis (European perch)	6 d	LC50	98	(Solbe et al., 1985)
Pimephales promelas (fathead minnow)	96 hrs	LOEC: Growth	58	(Barron and Adelman, 1985)
Rutilus rutilus (intersex roach)	7 d	LC50	108	(Solbe et al., 1985)
Salmo salar Ouananiche (landlocked Atlantic salmon)	12 d	LOEC (only conc. tested): Plasma vitellogenin	<4.8	(Ruby et al., 1987)
Salmo salar Ouananiche (landlocked Atlantic	12 d	NOEC (only conc. tested): Liver	>4.8	(Ruby et al., 1987)

Table 5a. Other Data on Effects of Cyanide on Freshwater Organisms.

Freshwater Species	Exposure Duration	Effect	Result (µg CN/L)	Reference
salmon)		vitellogenin		
Salmo salar Ouananiche (landlocked Atlantic salmon)	12 d	NOEC (only conc. tested): Hepatosomatic index	>4.8	(Ruby et al., 1987)
Salmo salar Ouananiche (landlocked Atlantic salmon)	12 d	NOEC (only conc. tested): Gonadosomatic index	>4.8	(Ruby et al., 1987)

*- in river water

LC50 = median lethal concentration

LT50 = time to 50% mortality

EC50 = median effect concentration

NOEC = no observed effect concentration

LOEC = lowest observed effect concentration

LC100 = lethal concentration to 100% of test population

Table 5b. Other Data on Effects of Cyanide on Saltwater Organisms.

Saltwater Species	Exposure Duration	Effect	Result (µg CN/L)	Reference
		Studies from Table 5 of the 1984 AWQC Document:		
Crassostrea sp. (oyster)	10 min	Suppressed ciliary activity	150	(Usuki, 1956)
Crassostrea sp. (oyster)	3 hrs	Inhibited ciliary activity	30,000	(Usuki, 1956)
Salmo salar (Atlantic salmon)	24 hrs	LC50	20-75	(Alabaster, 1983)
Lagodon rhomboides (pinfish)	24 hrs	LC50	69	(Daugherty and Garrett, 1951)
		Studies Published After the 1984 AWQC Document:		
Acartia tonsa (copepod)	48 hrs	LC50	195.0	(Parametrix, 2006)
Acartia tonsa (copepod)	48 hrs	LC50	191.2	(Parametrix, 2006)
Acartia tonsa (copepod)	48 hrs	LC50	217.1	(Parametrix, 2006)
Acartia tonsa (copepod)	48 hrs	LC50	234.7	(Parametrix, 2006)
Haliotis varia (abalone)	96 hrs	LOEC: Mortality	<400	(Lasut,1998)
Haliotis varia (abalone)	96 hrs	LOEC: Mortality	<1,062	(Lasut, 1998)
Plesiastrea versipora (coral)	3 hrs	NOEC-LOEC: Fluorescence	26-260	(Jones and Hoegh-Guldberg, 1999)
Plesiastrea versipora (coral)	3 hrs	LC100	260,136	(Jones and Hoegh-Guldberg, 1999)
Plesiastrea versipora (coral)	3 hrs	NOEC-LOEC: # of zooanthellae	26-260	(Jones and Hoegh-Guldberg, 1999)
Pocillopora damicornis (coral)	30 min	Zooxanthellae density	5,203	(Jones and Steven, 1997)
Stylophora pistillata (coral)	10 sec	LOEC: Photosynthesis of symbiotic algae	520,272	(Jones et al., 1999)
Scapharca inaequivalvis (clam)	11.2 d	LT50	26,000	(De Zwaan et al., 1993)
Dascyllus aruanus (damselfish)	120 sec	LOEC: Mortality	25,000	(Hanawa, et al., 1998)
Mugil auratus (golden mullet)	24 hrs	NOEC: Plasma parameters	159	(Ozretic & Krajnovic-Ozretic, 1993)

LT50 = time to 50% mortality; NOEC = no observed effect concentration; LOEC = lowest observed effect concentration; LC100 = lethal concentration to 100% of test population

REFERENCES

Abel, P.D.; Garner, S.M. (1986) Comparisons of median survival times and median lethal exposure times for *Grammarus pulex* exposed to cadmium, permethrin and cyanide, *Water Res.*, **20,** (5), 579.

Abram, F.S.H. (1964) An application of harmonics to fish toxicology, *Int. Jour. Air Water Pollut.*, **8,** 325.

Academy of Natural Sciences (1960) *The sensistivity of aquatic life to certain chemicals commonly found in industrial wastes,* Academy of Natural Sciences, Philadelphia, PA.

Alabaster, J.S., et al., (1983) The acute lethal toxicity of mixtures of cyanide and ammonia to smolts of salmon, *Salmo salar* L. at low concentrations of dissolved oxygen, *Jour. Fish Biol.*, **22,** 215.

Alazemi, B.M.; Lewis, J.W.; Andrews, E.B. (1996) Gill damage in the freshwater fish *Gnathonemus petersii* (Family: Mormyridae) exposed to selected pollutants: an ultrastrucutural study, *Environ. Technol.*, **17,** (3), 225.

American Public Health Association (1971) Method 207 Cyanide, in *Standard Methods for the Examination of Water and Wastewater, 13th Edition*, Taras, M.J.; Greenberg, A.E.; Hoak, R.D.; Rand, M.C., Eds., American Public Health Association, American Water Works Association, and Water Environment Federation, Washington, DC.

American Public Health Association (1998) Method 4500-CN Cyanide, in *Standard Methods for the Examination of Water and Wastewater, 20th Edition*, Clesceri, L.S.; Greenberg, A.E.; Eaton, A.D., Eds., American Public Health Association, American Water Works Association, and Water Environment Federation, Washington, DC.

Anderson, B.G. (1946) The toxicity thresholds of various sodium salts determined by the use of *Daphnia magna, Sewage Works J.*, **18,** (1), 82.

Anderson, P.D.; Weber, L.J. (1975) Toxic response as a quantitative function of body size, *Toxicol. Appl. Pharm.*, **33,** 471.

Asfaw, A.; Ellersieck, M.R.; Mayer, F.L. (2003) *Interspecies correlation estimations (ICE) for acute toxicity to aquatic organisms and wildlife. II. User manual and software*; Report No. EPA/600/R-03/106, U.S. Environmental Protection Agency, Washington, D.C.

ASTM (1998a) ASTM Designation D 4282-95. Standard Test Method for Determination of Free Cyanide in Water and Wastewater by Microdiffusion, in *Annual Book of ASTM Standards*, Vol. 14.02, ASTM, Philadelphia, PA.

ASTM (1998b) Standard test method for determination of free cyanide in water and wastewater by microdiffusion, ASTM D4282-95, in *Annual Book of ASTM Standards*, American Society for Testing and Materials, Philadelphia, PA.

Barron, M.G.; Adelman, I.R. (1985) Temporal characterization of growth of fathead minnow *(Pimephales promelas)* larvae during sublethal hydrogen cyanide exposure, *Comp. Biochem. Phys. C*, **81,** (2), 341.

Billard, R.; Roubaud, P. (1985) The effect of metals and cyanide on fertilization in rainbow trout (*Salmo Gardneri*). *Water Res.*, **19,** (2), 209.

Bills, T.D.; Marking, L.L. (1988) Control of nuisance populations of crayfish with traps and toxicants, *Prog. Fish Cult.*, **50,** 103.

Bills, T.D.; Marking, L.L.; Olson, L.E. (1977) Effects of residues of the polychlorinated biphenyl Aroclor 1254 on the sensitivity of rainbow trout to selected environmental contaminants, *Prog. Fish Cult.*, **39,** 150.

Black, H.H., et al. (1957) Industrial wastes guide -- by-product coke industry, *Sew. Ind.*, **29,** 53.

Bridges, W.R. (1958) *Sodium cyanide as a fish poison. Special Scientific Report - Fisheries No. 253*; Report No. U.S. Fish & Wildlife Service, Washington D.C.

Bringmann, G. (1975) Determination of the biologically harmful effect of water pollutants by means of the retardation if cell proliferation of the blue algae *Microcystis*., *Gesundheits-Ing.*, **96,** 238.

Bringmann, G.; Kuhn, R. (1959a) The toxic effects of waste water on aquatic bacteria, algae, and small crustaceans., *Gesundheits-Ing.*, **80,** 115.

Bringmann, G.; Kuhn, R. (1959b) Water toxicology studies with protozoans as test organisms., *Gesundheits-Ing.*, **80,** 239.

Bringmann, G.; Kuhn, R. (1976) Comparative results of the damaging effects of water pollutants against bacteria (*Pseudomonas putida*) and blue algae (*Microcystis aeruginosa*), *Gas-Wasserfach, Wasser-Abwasser*, **117,** 410.

Bringmann, G.; Kuhn, R. (1977a) Limiting values for the damaging action at water pollutants to bacteria (*Pseudomonas putida*) and green algae (*Scenedesmus quadricauda*) in the cell multiplication inhibition test., *Z. Wasser Abwasser Forsch.*, **10,** 87.

Bringmann, G.; Kuhn, R. (1977b) Results of the damaging effect of water pollutants on *Daphnia magna*, *Z. Wasser Abwasser Forsch.*, **10,** 161.

Bringmann, G.; Kuhn, R. (1978a) Limiting values for the noxious effects of water pollutant material to blue algae (*Microcystis aeruginosa*) and green algae (*Scenedesmus quadricauda*) in cell propagation inhibition tests, *Vom Wasser*, **100,** 249.

Bringmann, G.; Kuhn, R. (1978b) Testing of substances for their toxicity threshold: model organisms *Microcystis (Diplocystis) aeruginosa* and *Scenedesmus quadricauda*, *Mitt. Int. Ver. Theor. Angew. Limnol.*, **21,** 275.

Bringmann, G.; Kuhn, R. (1979) Comparison of toxic limiting concentrations of water contaminations toward bacteria, algae, and protozoa in the cell-growth inhibition test, *Jaistecj. Bauphys. Umwelttech*, **100,** 249.

Bringmann, G.; Kuhn, R. (1980a) Comparison of the toxicity threshold of water pollutants to bacteria, algae, and protozoa in the cell multiplication inhibition test, *Water Res.*, **14,** 231.

Bringmann, G.; Kuhn, R. (1980b) Determination of the harmful biological effect of water pollutants on protozoa. II. bacteriovorous ciliates, *Z. Wasser Abwasser Forsch.*, **13,** 26.

Bringmann, G.; Kuhn, R. (1981) Comparison of the effects of harmful substances on flagellates as well as ciliates and on halozoic bacteriophagous and saprozoic protozoa, *Gas-Wasserfach, Wasser-Abwasser*, **122,** 308.

Bringmann, G.; Kuhn, R. (1982) Results of toxic action of water pollutants on *Daphnia magna* Straus tested by an improved standardized procedure, *Z. Wasser Abwasser Forsch.*, **15,** 1.

Brix, K.V.; Cardwell, R.D.; Henderson, D.G.; Marsden, A.R. (2000) Site-specific marine water-quality criterion for cyanide, *Environ. Toxicol. Chem.*, **19,** (9), 2323.

Brockway, D.L. (1963) Some effects of sub-lethal levels of pentachlorophenol and cyanide of the physiology and behavior of a cichlid fish, *Cichlasoma bimaculatum* (linnaeus). M.S. Thesis. Oregon State University, Corvallis, Oregon.

Broderius, S.J. (1970) Determination of molecular hydrocyanic acid in water and studies of the chemistry and toxicity to fish of the nickelocyanide complex. Masters thesis, Oregon State University, Corvallis, OR.

Broderius, S.J. (1973) Determination of molecular hydrocyanic acid in water and studies of the chemistry and toxicity in fish of metal-cyanide complexes. Ph.D. Thesis, Oregon State University, Corvallis.

Broderius, S.J.; Lloyd, L.L.; Lind, D.T. (1977) Relative toxicity of free cyanide and dissolved sulfide forms to the fathead minnow (*Pimephales promelas*), *J. Fish. Res. Board Can.*, **34,** 2323.

Brown, V.M. (1968) The calculation of the acute toxicity of mixtures of poisons to rainbow trout, *Water Res.*, **2,** 723.

Buikema, A.L., Jr., et al. (1977) *Rotifer sensitivity to combinations of inorganic water pollutants. Bulletin 92*; Report No. Virginia Water Resources Research Center, Blacksburg, VA.

Burdick, G.E., et al. (1958) Toxicity of cyanide to brown trout and smallmouth bass, *New York Fish Game Jour.*, **5,** 133.

Cairns, J.; Buikema, A.L.; Heath, A.G.; Parker, B.C. (1978) *Effects of temperature on aquatic organism sensitivity to selected chemicals,* Bulletin 106, Virginia Water Resources Research Center, Blacksburg, VA.

Cairns, J., Jr.; Scheier, A. (1959) The relationship of bluegill sunfish body size to tolerance for some common chemicals, *Proc. 13th Ind. Waste Conf. Eng. Bull.*, **43,** 243.

Cairns, J., Jr., et al. (1965) A comparison of the sensitivity to certain chemicals of adult zebra danios *Brachydanio rerio* (Hamilton-Buchanan) and zebra dania eggs with that of adult bluegill sunfish *Lepomis macrochirus*, *Raf. Notulae Naturae. No. 381.*

Cairns, J., Jr., et al. (1976) Invertebrate response to thermal shock following exposure to acutely sub-lethal concentrations of chemicals, *Arch. Hydrobiol.*, **77,** 164.

Cairns, J., Jr.,; Scheier, A. (1958) The effect of periodic low oxygen upon toxicity of various chemicals to aquatic organisms. Proceedings of the 12th Industrial Waste Conference, Purdue University, Engineering Extension Series No. 94, *Engineering Bulletin*, **42,** 165.

Cairns, J., Jr.,; Scheier, A. (1968) A comparison of the toxicity of some common industrial waste components tested individually and combined, *Prog. Fish Cult.*, **30,** (1), 3.

Cairns, J.; Scheier, A. (1963) Environmental effects upon cyanide toxicity to fish, *Notulae Naturae*, **361,** 1.

Call, D.J.; Brooke, L.T. (1982) *Memorandum to Richard E. Siefert,* University of Wisconsin, Superior, WI.

Call, D.J.; Brooke, L.T.; Ahmad, N.; Richter, J.E. (1983) *Toxicity and Metabolism Studies with EPA Priority Pollutants and Related Chemicals in Freshwater Organisms*; Report No. PB83-263665, National Technical Information Service, Springfield, VA.

Call, D.J., et al. (1979) *Toxicity, bioconcentration, and metabolism of selected chemicals in aquatic organisms.*; Report No. Third Quarterly Progress Report. University of Wisconsin-Superior, Superior, Wisconsin.

Calleja, M.C.; Persoone, G.; Geladi, P. (1994) Comparative acute toxicity of the first 50 multicentre evaluation of *in vitro* cytotoxicity chemicals to aquatic non-vertebrates, *Arch. Environ. Con. Tox.*, **26,** (1), 69.

Carballo, M.; Torroba, M.; Munoz, M.J.; Sanchez, C.; Tarazona, J.V.; Dominguez, J. (1992) Effect of copper and cyanide on some immunological parameters and stress in rainbow trout (*Oncorhynchus mykiss*), *Fish Shellfish Immun.*, **2,** (2), 121.

Cardin, J.A. (1980) *Memorandum to John H. Gentile*; Report No. U.S. EPA, Narragansett, RI.

Cardwell, R.D.; Foreman, D.G.; Payne, T.R.; Wilbur, D.J. (1976) *Acute toxicity of selected toxicants to six species of fish*; Report No. EPA-600/3-76-008, U.S. Environmental Protection Agency, Washington, D.C.

Carr, R.S.; Curran, M.D.; Mazurkiewicz, M. (1986) Evaluation of the archiannelid *Dinophilus gyrociliatus* for use in short-term life-cycle toxicity tests, *Environ. Toxicol. Chem.*, **5,** 703.

Carter, L. (1962) Bioassay of trade wastes, *Nature*, **196,** 1304.

Chen, C.W.; Selleck, R.E. (1969) A kinetic model of fish toxicity threshold, *Jour. Water Pollut. Control Fed.*, **41,** R294.

Cheng, S.K.; Ruby, S.M. (1981) Effects of pulse exposure to sublethal levels of hydrogen cyanide on reproduction of American flagfish, *Arch. Environ. Contam. Toxicol.*, **10,** 105.

Clark, J.M.; Cardwell, R.D.; Gensemer, R.W. (2006) Toxicity of cyanide to aquatic-dependent wildlife, in *Cyanide in Water and Soil: Chemistry, Risk and Management*, Dzombak, D.A.; Ghosh, R.S.; Wong-Chong, G.M., Eds., Taylor and Francis/CRC Press, Boca Raton, FL.

Conway, C.H. (1952) Life history of the water shrew (*Sorex palustris navigator*), *The American Midland Naturalist*, **48,** (1), 219.

Costa, H.D.; Ruby, S.M. (1984) The effect of sublethal cyanide on vitellogenic parameters in rainbow trout *Salmo gairdneri*, *Arch. Environ. Con. Tox.*, **13,** 101.

Costa, H.H. (1965a) Responses of freshwater animals to sodium cyanide solutions. III. tadpoles of *Rana temporaria*, *Ceylon Jour. Sci.*, **5,** 97.

Costa, H.H. (1965b) Responses of freshwater animals to sodium cyanide solutions. I. fish, *Ceylon Jour. Sci.*, **5,** 41.

Costa, H.H. (1965c) Responses of freshwater animals to sodium cyanide solutions. II. *Gammarus pulex*, *Ceylon Jour. Sci.*, **5,** 88.

Costa, H.H. (1966) The effect of exercise on the survival of *Phoxinus phoxinus* L. in sodium cyanide solutions, *Hydrobiologia*, **28,** 241.

Daugherty, F.M.J.; Garrett, J.T. (1951) Toxicity levels of hydrocyanic acid and some industrial by-products, *Texas Jour. Sci.*, **3,** 391.

De Zwaan, A.; Cattan, O.; Putzer, V.M. (1993) Sulfide and cyanide induced mortality and anaerobic metabolism in the arcid blood clam *Scapharca inaequivalvis*, *Comp. Biochem. Phys. C*, **105,** (1), 49.

DeForest, D.K.; Giddings, J.M.; Gensemer, R.W. (2007) AWQC for cyanide: Protectiveness of threatened and endangered species, *Integrated Environmental Assessment and Management*, In prep.

Department of Scientific and Industrial Research (1956) *Water Pollution Research, 1955,* London.

Di Toro, D.M.; Allen, H.E.; Bergman, H.L.; Meyer, J.S.; Paquin, P.R.; Santore, R.C. (2001) Biotic ligand model of the acute toxicity of metals. 1. Technical basis, *Environ. Toxicol. Chem.*, **20,** (10), 2382.

Dixon, D.G.; Hodson, P.V.; Kaiser, K.L.E. (1987) Serum sorbitol dehydrogenase activity as an indicator of chemically induced liver damage in rainbow trout, *Environ. Toxicol. Chem.*, **6,** (9), 685.

Dixon, D.G.; Leduc, G. (1981) Chronic cyanide poisoning of rainbow trout and its effects on growth, respiration, and liver histopathology, *Arch. Environ. Contam. Toxicol.*, **10,** 117.

Dixon, D.G.; Sprague, J.B. (1981) Acclimation-induced changes in toxicity of arsenic and cyanide to rainbow trout, *Salmo gairdneri* Richardson, *Jour. Fish Biol.*, **18,** 579.

Doudoroff, P. (1956) Some experiments on the toxicity of complex cyanides to fish, *Fish Tox.*, **28,** (8), 1020.

Doudoroff, P. (1976) *Toxicity to fish of cyanides and related compounds*; Report No. EPA-600/3-76-038, U.S. Environmental Protection Agency, Cincinnati, OH.

Doudoroff, P.; Leduc, G.; Schneider, C.R. (1966) Acute toxicity to fish of solutions containing complex metal cyanides, in relation to concentrations of molecular hydrocyanic acid, *T. Am. Fish. Soc.*, **95,** 6.

Dowden, B.F.; Bennett, H.J. (1965) Toxicity of Selected Chemicals to Certain Animals, *J. Water Pollut. Control Fed.*, **37,** (9), 1308.

Downing, K.M. (1954) The influence of dissolved oxygen on the toxicity of potassium cyanide to rainbow trout, *Jour. Exp. Biol.*, **31,** 161.

Dunning, J.B., Jr., (1993) *CRC handbook of avian body masses,* CRC Press, Boca Raton, FL.

Dwyer, F.J.; Mayer, F.L.; Sappington, L.C.; Buckler, D.R.; Bridges, C.M.; Greer, I.E.; Hardesty, D.K.; Henke, C.E.; Ingersoll, C.G.; Kunz, J.L.; Whites, D.W.; Augspurger, T.; Mount, D.R.; Hattala, K.; Neuderfer, G.N. (2005) Assessing contaminant sensitivity of endangered and threatened species: Part I. Acute toxicity of five chemicals, *Arch. Environ. Con. Tox.*, **48,** 143.

Ebbs, S.D.; Wong-Chong, G.M.; Bond, B.S.; Bushey, J.T.; Neuhauser, E.F. (2006) Biological transformation of cyanide in water and soil, in *Cyanide in Water and Soil: Chemistry, Risk and Management*, Dzombak, D.A.; Ghosh, R.S.; Wong-Chong, G.M., Eds., Taylor and Francis/CRC Press, Boca Raton, FL.

Eisler, R. (1991) *Cyanide hazards to fish, wildlife, and invertebrates: A synoptic review, Biological Report 85(1.23),* U.S. Fish and Wildlife Service, Washington, D.C.

ENSR (2005a) *Acute Toxicity of Cyanide to the Frog, Rana berlandieri, in Horsetooth Reservoir Water Under Flow-through Conditions*; Report No. 8503-124-020-076, ENSR Corporation, Fort Collins, CO.

ENSR (2005b) *Acute Toxicity of Cyanide to the Frog, Rana pipiens, in Horsetooth Reservoir Water Under Flow-through Conditions*; Report No. 8503-124-020-075, ENSR Corporation, Fort Collins, CO.

ENSR (2005c) *Acute Toxicity of Cyanide to the Frog, Xenopus laevis, in Horsetooth Reservoir Water Under Flow-through Conditions*; Report No. 8503-124-020-074, ENSR Corporation, Fort Collins, CO.

European Commission (1996) *Technical guidance document in support of commission directive 93/67/EEC on risk assessment for new notified substances and commission regulation (EC) No. 1488/94 on risk assessment for existing substances,* Office of Official Publication of the European Communities, Brussels, Luxembourg.

Ewell, W.S.; Gorsuch, J.W.; Kringle, R.O.; Robillard, K.A.; Spiegel, R.C. (1986) Simultaneous evaluation of the acute effects of chemical on seven aquatic species, *Environ. Toxicol. Chem.*, **5,** 831.

Fitzgerald, G.P.; Gerloff, G.C.; Skoog, F. (1952) Studies on chemicals with selective toxicity to blue-green algae, *Sewage Ind. Wastes*, **24,** 888.

Gardner, G.; Berry, W. (1981) *Memorandum to John H. Gentile,* U.S. Environmental Protection Agency, Narragansett, RI.

Gardner, G.; Nelson, W. (1981) *Memorandum to John H. Gentile,* U.S. Environmental Protection Agency, Narragansett, RI.

Gensemer, R.W.; DeForest, D.; Coyner, A.; Clark, J.; Cardwell, R.D.; Dzombak, D.A.; Higgins, C.; Santore, R.C. (2003) *Reassessment of cyanide criteria for the protection of aquatic life and wildlife: Phase I annual progress report (01-ECO-1),* Water Environment Research Foundation, Alexandria, VA.

Gensemer, R.W.; DeForest, D.K.; Stenhouse, A.J.; Higgins, C.J.; Cardwell, R.D. (2006) Aquatic toxicity of cyanide, in *Cyanide in Water and Soil: Chemistry, Risk and Management*, Dzombak, D.A.; Ghosh, R.S.; Wong-Chong, G.M., Eds., Taylor and Francis/CRC Press, Boca Raton, FL.

Gentile, S.L. (1980) *Memorandum to John H. Gentile,* U.S. Environmental Protection Agency, Narragansett, RI.

Ghosh, R.S.; Dzombak, D.A.; Drop, S.M.; Zheng, A. (2006) Analysis of cyanide in water, in *Cyanide in Water and Soil: Chemistry, Risk and Management*, Dzombak, D.A.; Ghosh, R.S.; Wong-Chong, G.M., Eds., Taylor and Francis/CRC Press, Boca Raton, FL.

Hanawa, M.; Harris, L.; Graham, M.; Farrell, A.P.; Bendell-Young, L.I. (1998) Effects of cyanide exposure on *Dascyllus aruanus*, a tropical marine fish species: lethality, anesthesia and physiological effects, *Aquarium Sci. Conservation*, **2,** (1), 21.

Hansen, D.J.; Burgess, R.M. (1987) *Changes to the acute values of some marine copepods in 1987 documents. Memorandum to Charles E. Stephan,* U.S. Environmental Protection Agency, Narragansett, RI.

Henderson, C.; Pickering, Q.H.; Lemke, A.E. (1961) The effect of some organic cyanides (nitriles) on fish. Proceedings of the 15th Industrial Waste Conference, Purdue University, Series No. 106, *Engineering Bulletin*, **65,** (2), 120.

Herbert, D.W.M.; Merkens, J.C. (1952) The toxicity of potassium cyanide to trout, *Journal of Experimental Biology*, **29,** 632.

Higgins, C.J.; Dzombak, D.A. (2006) Free cyanide sorption on freshwater sediment and model components, *Soil Sediment Contam.*, **15,** (5), 497.

Ishio, S. (1965) Behavior of fish exposed to toxic substances, in *Advances in Water Pollution Research*, Jaag, O., Ed., Pergamon Press, New York.

Johns, M.J.; Gentile, J.S. (1981) *Results of acute toxicity tests conducted with cyanide at ERL, Memorandum to Dave Hansen,* U.S. Environmental Protection Agency, Narragansett, RI.

Jones, J.R.E. (1941) A study of the relative toxicity of anions, with *Polycelis nigra* as test animals, *Jour. Exp. Biol.*, **18,** 170.

Jones, J.R.E. (1947) The oxygen consumption of *Gasterosteus aculeatus* L. in toxic solutions, *Jour. Exp. Biol.*, **23,** 298.

Jones, R.J.; Hoegh-Guldberg, O. (1999) Effects of cyanide on coral photosynthesis: implications for identifying the cause of coral bleaching and for assessing the environmental effects of cyanide, *Mar. Ecol-Prog. Ser.*, **177,** 83.

Jones, R.J.; Kildea, T.; Hoegh-Guldberg, O. (1999) PAM chlorophyll fluorometry: a new in situ technique for stress assessment in Scleractinian corals, used to examine the effects of cyanide from cyanide fishing., *Mar. Pollut. Bull.*, **38,** (10), 864.

Jones, R.J.; Steven, A.L. (1997) Effects of cyanide on corals in relation to cyanide fishing on reefs, *Mar. Freshwater Res.*, **48,** (517-522).

Karsten, A. (1934) Investigations of the effect of cyanide on Black Hills trout, *Black Hills Eng.*, **22,** 145.

Kavanaugh, M.C.; Deeb, R.A.; Markowitz, D.; Dzombak, D.A.; Zheng, A.; Theis, T.L.; Young, T.C.; Luthy, R.G. (2003) *Cyanide formation and fate in complex effluents and its relation to water quality criteria*; Report No. 98-HHE-5, Water Environment Research Foundation, Alexandria, VA.

Kimball, G.L.; Smith, L.L.; Broderius, S.J. (1978) Chronic toxicity of hydrogen cyanide to the bluegill, *T. Am. Fish. Soc.*, **107,** (2), 341.

Koenst, W.M.; Smith, L.L.; Broderius, S.J. (1977) Effect of chronic exposure of brook trout to sublethal concentrations of hydrogen cyanide, *Environ. Sci. Technol.*, **11,** (9), 883.

Kondo, T.; Tsudzuki, T. (1980) Energy supply for potassium uptake rhythm in a duckweed *Lemna gibba* G-3, *Plant Cell Physiol.*, **21,** 433.

Köster, H.W. (2001) *Risk assessment of historical soil contamination with cyanides; origin, potential human exposure and evaluation of intervention values.*; Report No. RIVM 711701019, National Institute of Public Health and the Environment,

Kovacs, T.G. (1979) The effect of temperature on cyanide toxicity to rainbow trout (*Salmo gairdneri*). M.Sc. Thesis, Concordia University, Montreal, Quebec.

Kovacs, T.G.; Leduc, G. (1982a) Acute toxicity of cyanide to rainbow trout (*Salmo gairdneri*) acclimated at different temperature, *Can. J. Fish. Aquat. Sci.*, **39,** 1426.

Kovacs, T.G.; Leduc, G. (1982b) Sublethal toxicity of cyanide to rainbow trout (*Salmo gairdneri*) at different temperatures, *Can. J. Fish. Aquat. Sci.*, **39,** 1389.

Lanno, R.P.; Dixon, D.G. (1996) The comparative chronic toxicity of thiocyanate and cyanide to rainbow trout., *Aquat. Toxicol.*, **36,** 177.

Lasut, M.T. Effects of salinity-cyanide interaction on the mortality of abalone *Haliotis varia* (Haliotidae: Gastropoda). presented at Workshop of the Tropical Marine Mollusc Programme (TMMP), Lombok (Indonesia), August 19-29, 1998.

Leduc, G. (1966) Some physiological and biochemical responses of fish to chronic poisoning by cyanide. Ph.D. thesis, Oregon State University, Corvallis, Oregon.

Leduc, G. (1978) Deleterious effects of cyanide on early life stages of Atlantic salmon (*Salmo salar*), *Jour. Fish Res. Board Can.*, **35,** 166.

Leduc, G.; Chan, K.K.S. The effects of chronic cyanide poisoning on the tolerance of rainbow trout to varying salinity, presented at Water Pollution Research in Canada, Proc. 10th Canadian Symp. Water Pollut. Res.: 1975.

Lee, D.R. (1976) Development of an invertebrate bioassay to screen petroleum refinery effluents discharged into freshwater. Ph.D. thesis, Virginia Polytechnic Institute and State University, Blacksburg, VA.

Lesniak, J.A. (1977) A histological approach to the study of sublethal cyanide effects on rainbow trout ovaries. M.S. Thesis. Concordia University, Montreal.

Lesniak, J.A.; Ruby, S.M. (1982) Histological and quantitative effects of sublethal cyanide exposure on oocyte development in rainbow trout, *Arch. Environ. Contam. Toxicol.*, **11,** 343.

Lewis, W.M.; Tarrant, R.M.J. (1960) Sodium cyanide in fish management and culture, *Prog. Fish-Cult.*, **22,** 177.

Lilius, H.; Hastbacka, T.; Isomaa, B. (1995) A comparison of the toxicity of 30 reference chemicals to *Daphnia magna* and *Daphnia pulex*, *Environ. Toxicol. Chem.*, **14,** (12), 2085.

Lilius, H.; Isomaa, B.; Holmstrom, T. (1994) A comparison of the toxicity of 50 reference chemicals to freshly isolated rainbow trout hepatocytes and *Daphnia magna*, *Aquat. Toxicol.*, **30,** 47.

Lind, D.T.; Smith, L.L.; Broderius, S.J. (1977) Chronic effects of hydrogen cyanide on the fathead minnow, *J. Water Pollut. Control Fed.*, **49,** 262.

Lipschuetz, M.; Cooper, A.L. (1955) Comparative toxicities of potassium cyanide and potassium coprocyanide to the western blacknosed dace (*Rhinichthys atratulus meleagris*), *New York Fish and Game Journal*, **2,** 194.

Lloyd, R.; Jordan, D.H.M. (1964) Predicted and observed toxicities of several sewage effluents to rainbow trout: a further study, *Inst. Sew. Purif., Jour. Proc. (Pt. 2)*, 183.

Lomte, V.S.; Jadhav, M.L. (1982) Effects of toxic compounds on oxygen consumption in the fresh water bivalve, *Corbicula regularis* (Prime, 1860), *Comp. Physiol. Ecol.*, **7,** 31.

Lund, B.L. (1918) The toxic action of KCN and its relation to the state of nutrition and age of the cell as shown by *Paramecium* and *Didinium*, *Biol. Bull.*, **35,** 211.

Lussier, S.M.; Cardin, J.A. (1985) *Results of acute toxicity tests conducted with cyanide at ERL, Narragansett. Memorandum to David Hansen,* U.S. Environmental Protection Agency, Narragansett, RI.

Lussier, S.M.; Gentile, J.H.; Walker, J. (1985) Acute and chronic effects of heavy metals and cyanide on *Mysidopsis bahia* (Crustacea:Mysidacea), *Aquat. Toxicol.*, **7,** (1-2), 25.

Marking, L.L.; Bills, T.D.; Crowther, J.R. (1984) Effects of five diets on sensitivity of rainbow trout to eleven chemicals, *Prog. Fish Cult.*, **46,** (1), 1.

McCracken, I.R.; Leduc, G. (1980) Allometric growth response of exercised rainbow trout to cyanide poisoning, in *Aquatic Toxicology. ASTM STP 707*, Eaton, J.G., et al.,, Ed., American Society for Testing and Materials, Philadelphia, Pennsylvania.

McGeachy, S.M.; Leduc, G. (1988) The influence of season and exercise on the lethal toxicity of cyanide to rainbow trout *Salmo gairdneri*, *Arch. Environ. Con. Tox.*, **17,** (3), 313.

Meeussen, J.L.; Keizer, M.G.; de Haan, F.A.M. (1992) Chemical stability and decomposition rate of iron cyanide complexes in soil solutions, *Environ. Sci. Technol.*, **26,** 511.

Moore, S.L.; Kin, S.R. (1968) Cyanide pollution and emergency duty: train wreck, Dunreith, Indiana. Proc. 23rd Ind. Waste Conf., Purdue Univ., Eng. Ext. Series No. 132 (Pt. 1), *Eng. Bull*, **53,** 583.

Morgan, W.S.G. (1979) Fish locomotor behavior patterns as a monitoring tool, *Jour. Water Pollut. Control Fed.*, **51,** 580.

Morgan, W.S.G.; Kuhn, P.C. (1974) A method to monitor the effects of toxicants upon breathing rates of largemouth bass (*Micropterus salmoides* Lacepede), *Water Res.*, **8,** 67.

Munoz, M.J.; Carballo, M.; Tarazona, J.V. (1991) The effect of sublethal levels of copper and cyanide on some biochemical parameters of rainbow trout along subacute exposition, *Comp. Biochem. Phys. C*, **100,** (3), 577.

Nalecz-Jawecki, G.; Sawicki, J. (1998) Toxicity of inorganic compounds in the spirotox test: a miniaturized version of the *Spirostomum ambiguum* test, *Arch. Environ. Con. Tox.*, **34,** (1), 1.

Negilski, D.S. (1973) Individual and combined effects of cyanide, pentachlorophenol and zinc on juvenile chinook salmon and invertebrates in model stream communities. Masters thesis, Oregon State University, Corvallis, OR.

Neil, J.H. Some effects of potassium cyanide on speckled trout (*Salvelinus fontinalis*), presented at 4th Ontario Industrial Waste Conference, Ontario Water Resources Commission: Toronto, 1957; p 74.

Nelson, E.B.; Tolbert, N.E. (1970) Glycolate dihydrogenase in green algae, *Arch. Biochem. Biophys.*, **141,** 102.

NMFS (2005) *Species protected under the Endangered Species Act, http://www.nmfs.noaa.gov/pr/species/esa_species.htm,* National Marine Fisheries Service, Silver Spring, MD.

Northwestern Aquatic Sciences (2005a) *Yellow rock crab (Cancer irroratus) larval acute toxicity test of cyanide*; Report No. 677-4, Northwestern Aquatic Sciences, Newport, OR.

Northwestern Aquatic Sciences (2005b) *Yellow rock crab (Cancer irroratus) larval acute toxicity test of cyanide*; Report No. 677-2, Northwestern Aquatic Sciences, Newport, OR.

Northwestern Aquatic Sciences (2005c) *Yellow rock crab (Cancer irroratus) larval acute toxicity test of cyanide*; Report No. 677-1, Northwestern Aquatic Sciences, Newport, OR.

Oseid, D.M.; Smith, L.L. (1979) The effects of hydrogen cyanide on *Aselius communis* and *Gammarus pseudolimnaeus* and changes in their competitive response when exposed simultaneously, *B. Environ. Contam. Tox.*, **21,** 439.

Ozretic, B.; Krajnovic-Ozretic, M. (1993) Plasma sorbitol dehydrogenase, glutamate dehydrogenase, and alkaline phosphatase as potential indicators of liver intoxication in grey mullet, *B. Environ. Contam. Tox.*, **50,** (4), 586.

Pablo, F.; Buckney, R.T.; Lim, R.P. (1997a) Toxicity of cyanide, iron-cyanide complexes, and a blast furnace effluent to larvae of the doughboy scallop, *Chlamys asperrimus*, *B. Environ. Contam. Tox.*, **58,** 93.

Pablo, F.; Stauber, J.L.; Buckney , R.T. (1997b) Toxicity of cyanide and cyanide complexes to the marine diatom *Nitzschia closterium*, *Water Res.*, **31,** (10), 2435.

Pablo, R.; Buckney, R.T.; Lim, R.P. (1997c) Toxicity of cyanide, iron-cyanide complexes, and a blast furnace effluent to the banana prawn, *Penaeus monodon*, *B. Environ. Contam. Tox.*, **58,** 822.

Parametrix (2005) *96-hr Acute Toxicity of Cyanide to Gasterosteus aculeatus Under Flow-through Conditions*; Report No. 3539-15, Parametrix Environmental Research Laboratory, Albany, Oregon.

Parametrix (2006) *96-hr Acute Toxicity of Cyanide to Acartia tonsa*; Report No. 3555-1-4, Parametrix Environmental Research Laboratory, Albany, Oregon.

Patrick, R.; Cairns, J.; Scheier, A. (1968) The relative sensitivity of diatoms, snails, and fish to twenty common constituents of industrial wastes, *Prog. Fish Cult.*, **30,** 137.

Pedroso, M.S.; Pinho, G.L.L.; Bianchini, A. (2007) Mechanism of acute silver toxicity in the euryhaline copepod, *Acartia tonsa. Aquat. Toxicol.,* **82**, 173.

Rause; Gasaway Alaska Department of Fish and Game, Notebook Series, Moose; http://www.adfg.state.ak.us/pubs/notebook/biggame/moose.php. http://www.adfg.state.ak.us/pubs/notebook/biggame/moose.php

Raymond, P.; Leduc, G. (1985) Investigation on the mode of action of cyanide by monitoring various physiological parameters in rainbow trout *Salmo gairdneri* exposed during 20 days to sublethal cyanide levels, *Canadian Technical Report Of Fisheries And Aquatic Sciences*, **1368,** 165.

Renn, C.E. (1955) Biological properties and behaviors of cyanogenic wastes, *Sew. Ind. Wastes*, **27,** 297.

Rippon, G.D.; LeGras, C.A.; Hyne, R.V.; Cusbert, P.J. (1992) *Toxic effects of cyanide on aquatic animals of the Aligator Rivers region,* Technical Memorandum 39, Australian Government Publishing Service, Canberra, Australia.

Roback, S.S. (1965) Environmental requirements of Trichoptera, in *Biological problems in water Pollution*, Tarzwell, C.M., Ed., Robert A. Taft Sanitary Engineering Center, Cincinnati, Ohio.

Ruby, S.M., et al. (1979) Inhibition of spermatogenesis in rainbow trout during chronic cyanide poisoning, *Arch. Environ. Con. Tox.*, **8,** 533.

Ruby, S.M.; Idler, D.R.; So, Y.P. (1986) The effect of sublethal cyanide exposure on plasma vitellogenin levels in rainbow trout (*Salmo gairdneri*) during early vitellogenesis, *Arch. Environ. Con. Tox.*, **15,** 603.

Ruby, S.M.; Idler, D.R.; So, Y.P. (1987) Changes in plasma, liver, and ovary vitellogenin in landlocked atlantic salmon following exposure to sublethal cyanide., *Arch. Environ. Con. Tox.*, **16,** (4), 507.

Ruby, S.M.; Jaroslawski, P.; Hull, R. (1993) Lead and cyanide toxicity in sexually maturing rainbow trout, *Oncorhynchus mykiss* during spermatogenesis, *Aquat. Toxicol.*, **26,** (3-4), 225.

Sanoli, B.A.; Kanabur, V.V. (2001) Acute toxicity of cyanide and formalin to a freshwater fish *Lepidocepalichithys guntea* (catfish), *Indian J. Fish.*, **48,** (1), 99.

Santore, R.C.; Di Toro, D.M.; Paquin, P.R.; Allen, H.E.; Meyer, J.S. (2001) Biotic ligand model of the acute toxicity of metals. 2. Application to acute copper toxicity in freshwater fish and *Daphnia*, *Environ. Toxicol. Chem.*, **20,** (10), 2397.

Santore, R.C.; Driscoll, C.T. (1995) The CHESS model for calculating chemical equilibria in soils and solutions, in *Chemical Equilibrium and Reaction Models*, Loeppert, R.H.; Schwab, A.P.; Goldberg, S., Eds., American Society of Agronomy, Madison, WI.

Sarkar, S. (1990) Toxicity evaluation of sodium cyanide to fish and aquatic organisms., *Sci. Cult.*, **56,** 165.

Schimmel, S. (1981) *Final report on toxicity of cyanide sheepshead minnow early life stages. Memorandum to John Gentile,* U.S. Environmental Protection Agency, Narragansett, RI.

Scott, K.J.; Yevich, P.P.; Boothman, W.S. (1983) *Toxicological methods using the benthic amphipod Ampelisca abdita Mills. Contribution 576,* U.S. Environmental Protection Agency, Environmental Research Laboratory, Narragansett, RI.

Shelford, V.E. (1917) An experimental study of the effects of gas waste upon fishes, with especial reference to stream pollution, *Bull. Illinois State Laboratory Nat. History Vol XI, Article VI*, 381.

Skibba, W.D. (1981) Trout test with *Salmo gardineri* Richardson for determining the acute toxicity of pollutants and test measurements for sodium cyanide, a cyanidic copper and azaplant, *Acta Hydroch. Hydrob.*, **9,** (1), 3.

Smith, L.L.; Broderius, S.J.; Oseid, D.M.; Kimball, G.L.; Koenst, W.M. (1978) Acute toxicity of hydrogen cyanide to freshwater fishes, *Arch. Environ. Con. Tox.*, **7,** 325.

Smith, L.L.; Broderius, S.J.; Oseid, D.M.; Kimball, G.L.; Koenst, W.M.; Lind, D.T. (1979) *Acute and Chronic Toxicity of HCN to Fish and Invertebrates*; Report No. EPA-600/3-79-009, U.S. Environmental Protection Agency, Washington, D.C.

Smith, M.J.; Heath, A.G. (1979) Acute toxicity of copper, chromate, zinc and cyanide to freshwater fish: effect of different temperatures, *Bull. Environ. Contam. Toxicol.*, **22,** 113.

Solbe, L.G.; Cooper, V.A.; Willis, C.A.; Mallett, M.J. (1985) Effects of pollutants in fresh waters on European non-salmonid fish I: Non-metals, *Fish Biol.*, **27,** (A), 197.

Speyer, M.R. (1975) Some effects of chronic combined arsenic and cyanide poisoning on the physiology of rainbow trout. M.Sc. Thesis, Concordia University, Montreal.

Stanley, R.A. (1974) Toxicity of heavy metals and salts to Eurasian watermilfoil (*Myriophyllum spicatum L.*), *Arch. Environ. Con. Tox.*, **2,** 331.

Steele, R.L.; Thursby, G.B. (1983) A toxicity test using life stages of *Champia parvula* (Rhodophyta). in *Aquatic Toxicology and Hazard Assessment: Sixth Symposium, STP 802*, Bishop, W.E.; Cardwell, R.D.; Heidolph, B.B., Eds., American Society for Testing and Materials, Philadelphia, PA.

Stephan, C.E. (2002) Use of species sensitivity distributions in the derivation of water quality criteria for aquatic life by the U.S. Environmental Protection Agency, in *Species sensitivity distributions in ecotoxicology*, Posthuma, L.; Suter, G.W., III; Traas, T.P., Eds., Lewis Publishers, Boca Raton, FL.

Summerfelt, R.C.; Lewis, W.M. (1967) Repulsion of green sunfish by certain chemicals, *Jour. Water Pollut. Control Fed.*, **39,** 2030.

Szabo, A.; Ruby, S.M.; Rogan, F.; Amit, Z. (1991) Changes in brain dopamine levels, oocyte growth and spermatogenesis in rainbow trout, *Oncorhynchus mykiss*, following sublethal cyanide exposure, *Arch. Environ. Con. Tox.*, **21,** (1), 152.

Tryland, O.; Grande, M. (1983) Removal of cyanide from scrubber effluents and its effect on toxicity to fish, *Vatten*, **39,** 168.

Turnbull, R., et al. (1954) Toxicity of various refinery materials to freshwater fish, *Ind. Eng. Chem.*, **46,** 324.

U.S. EPA (1984) *Guidelines for deriving numerical aquatic site-specific water quality criteria by modifying national criteria*; Report No. PB85-121101, U.S. Environmental Protection Agency, Washington, D.C.

U.S. EPA (1985a) *Ambient water quality criteria for cyanide - 1984*; Report No. EPA 440/5-84-028, U.S. Environmental Protection Agency, Washington, D.C.

U.S. EPA (1985b) *Guidelines for deriving numerical national water quality criteria for the protection of aquatic organisms and their uses*; Report No. PB85-227049, U.S. Environmental Protection Agency, Washington, DC.

U.S. EPA (1988) *Calculation of the final acute value for water quality criteria for aquatic organisms*; Report No. EPA/600/3-88/018, U.S. Environmental Protection Agency, Washington, D.C.

U.S. EPA (1993a) *Great Lakes water quality initiative criteria documents for the protection of wildlife (proposed); DDT, Mercury, 2,3,7,8-TCDD, PCBs. EPA-822-R-93-007,* U.S. Environmental Protection Agency, Washington, DC.

U.S. EPA (1993b) *Wildlife exposure factors handbook*; Report No. EPA/600/R-93/187, U.S. Environmental Protection Agency, Office of Research and Development, Washington D.C.

U.S. EPA (1994) *Interim guidance on determination and use of water-effect ratios for metals*; Report No. EPA-823-B-94-001, U.S. Environmental Protection Agency, Washington, D.C.

U.S. EPA (1995) Final water quality guidance for the Great Lakes system; final rule, *Federal Register*, **60,** (56), 15366.

U.S. EPA (1999) *1999 update of ambient water quality criteria for ammonia*; Report No. EPA-822-R-99-014, U.S. Environmental Protection Agency, Washington, DC.

U.S. EPA (2002) *National recommended water quality criteria: 2002*; Report No. EPA-822-R-02-047, U.S. Environmental Protection Agency, Washington, DC.

U.S. EPA (2003a) *2003 Draft update of ambient water quality criteria for copper*; Report No. EPA 822-R-03-026, U.S. Environmental Protection Agency, Washington, D.C.

U.S. EPA (2003b) *Draft strategy: Proposed revisions to the "guidelines for deriving numerical national water quality criteria for the protection of aquatic organisms and their uses",* U.S. Environmental Protection Agency, Washington, D.C.

U.S. EPA (2003c) *Multimedia, multipathway, and multireceptor risk assessment (3MRA)*; Report No. EPA530-D-03-001b, U.S. Environmental Protection Agency, ORD, Washington, D.C.

U.S. EPA (2003d) *Procedures for the derivation of equilibrium partitioning sediment benchmarks (ESBs) for the protection of benthic organisms: PAH mixtures*; Report No. EPA-600-R-02-013, U.S. Environmental Protection Agency, Washington, D.C.

U.S. EPA; USFWS; NOAA (2001) Memorandum of agreement between the Environmental Protection Agency, Fish and Wildlife Service and National Marine Fisheries Service regarding enhanced coordination under the Clean Water Act and the Endangered Species Act, *Federal Register*, **66,** (36), 11202.

Usuki, I. (1956) A comparison of the effects of cyanide and azide on the ciliary activity of the oyster gill, *Sci, Rept. Tohoku University, Fourth Sci.*, **22,** 137.

Wallen, I.E.; Greer, W.C.; Lasater, R. (1957) Toxicity to *Gambusia affinis* of certain pure chemicals in turbid waters, *Sewage Ind. Wastes*, **29,** 695.

Washburn, G.N. (1948) The toxicity to warm-water fishes of certain cyanide plating and carburising salts before and after treatment by the alkali-chlorination method, *Sew. Works Jour.*, **20,** 1074.

Webster, D.A.; Hackett, D.P. (1965) Respiratory chain of colorless algae. I. Chlorophyta and Euglenophyta., *Plant Physiol.*, **40,** 1091.

Whittingham, C.P. (1952) Inhibition of photosynthesis by cyanide, *Nature*, **169,** 838.

Woker, H.; Wuhrmann, K. (1950) Contributions to fish toxicology. VI. the sensitivity of different species of fish to ammonia, hydrocyanic acid, and phenol, *Rev. Suisse Zool.*, **57,** 548.

Zheng, A.; Dzombak, D.A.; Luthy, R.G.; Sawyer, B.; Lazouskas, W.; Tata, O.; Delaney, M.F.; Zilitinkevitch, L.; Sebroski, J.R.; Swartling, R.S.; Drop, S.; Flaherty, J. (2003) Evaluation and testing of analytical methods for cyanide species in municipal and industrial contaminated waters, *Environ. Sci. Technol.*, **37,** (1), 107.

WASTEWATER UTILITY

Alabama
Montgomery Water Works & Sanitary Sewer Board

Alaska
Anchorage Water & Wastewater Utility

Arizona
Glendale, City of, Utilities Department
Mesa, City of
Peoria, City of
Phoenix Water Services Dept.
Pima County Wastewater Management
Safford, City of

Arkansas
Little Rock Wastewater Utility

California
Central Contra Costa Sanitary District
Corona, City of
Crestline Sanitation District
Delta Diablo Sanitation District
Dublin San Ramon Services District
East Bay Dischargers Authority
East Bay Municipal Utility District
Eastern Municipal Water District
El Dorado Irrigation District
Fairfield-Suisun Sewer District
Fresno Department of Public Utilities
Inland Empire Utilities Agency
Irvine Ranch Water District
Las Virgenes Municipal Water District
Livermore, City of
Los Angeles, City of
Los Angeles County, Sanitation Districts of
Napa Sanitation District
Orange County Sanitation District
Palo Alto, City of
Riverside, City of
Sacramento Regional County Sanitation District
San Diego Metropolitan Wastewater Department, City of
San Francisco, City & County of
San Jose, City of
Santa Barbara, City of
Santa Cruz, City of
Santa Rosa, City of
South Bayside System Authority
South Coast Water District
South Orange County Wastewater Authority
Stege Sanitary District
Sunnyvale, City of
Union Sanitary District
West Valley Sanitation District

Colorado
Aurora, City of
Boulder, City of
Greeley, City of
Littleton/Englewood Water Pollution Control Plant
Metro Wastewater Reclamation District, Denver

Connecticut
Greater New Haven WPCA
Stamford, City of

District of Columbia
District of Columbia Water & Sewer Authority

Florida
Broward, County of
Fort Lauderdale, City of
Miami-Dade Water & Sewer Authority
Orange County Utilities Department
Reedy Creek Improvement District
Seminole County Environmental Services
St. Petersburg, City of
Tallahassee, City of
Tampa, City of
Toho Water Authority
West Palm Beach, City of

Georgia
Atlanta Department of Watershed Management
Augusta, City of
Clayton County Water Authority
Cobb County Water System
Columbus Water Works
Fulton County
Gwinnett County Department of Public Utilities
Savannah, City of

Hawaii
Honolulu, City & County of

Idaho
Boise, City of

Illinois
American Bottoms Wastewater Treatment Plant
Greater Peoria Sanitary District
Kankakee River Metropolitan Agency
Metropolitan Water Reclamation District of Greater Chicago
Wheaton Sanitary District

Iowa
Ames, City of
Cedar Rapids Wastewater Facility
Des Moines, City of
Iowa City

Kansas
Johnson County Unified Wastewater Districts
Unified Government of Wyandotte County/ Kansas City, City of

Kentucky
Louisville & Jefferson County Metropolitan Sewer District
Sanitation District No. 1

Louisiana
Sewerage & Water Board of New Orleans

Maine
Bangor, City of
Portland Water District

Maryland
Anne Arundel County Bureau of Utility Operations
Howard County Department of Public Works
Washington Suburban Sanitary Commission

Massachusetts
Boston Water & Sewer Commission
Massachusetts Water Resources Authority (MWRA)
Upper Blackstone Water Pollution Abatement District

Michigan
Ann Arbor, City of
Detroit, City of
Holland Board of Public Works
Saginaw, City of
Wayne County Department of Environment
Wyoming, City of

Minnesota
Rochester, City of
Western Lake Superior Sanitary District

Missouri
Independence, City of
Kansas City Missouri Water Services Department
Little Blue Valley Sewer District
Metropolitan St. Louis Sewer District

Nebraska
Lincoln Wastewater System

Nevada
Henderson, City of
Reno, City of

New Jersey
Bergen County Utilities Authority
Ocean County Utilities Authority
Passaic Valley Sewerage Commissioners

New York
New York City Department of Environmental Protection

North Carolina
Charlotte/Mecklenburg Utilities
Durham, City of
Metropolitan Sewerage District of Buncombe County
Orange Water & Sewer Authority

Ohio
Akron, City of
Butler County Department of Environmental Services
Columbus, City of
Metropolitan Sewer District of Greater Cincinnati
Northeast Ohio Regional Sewer District
Summit, County of

Oklahoma
Oklahoma City Water & Wastewater Utility Department
Tulsa, City of

Oregon
Albany, City of
Clean Water Services
Eugene, City of
Gresham, City of
Portland, City of Bureau of Environmental Services
Water Environment Services

Pennsylvania
Philadelphia, City of
University Area Joint Authority

South Carolina
Charleston Water System
Mount Pleasant Waterworks & Sewer Commission
Spartanburg Water

Tennessee
Cleveland, City of
Knoxville Utilities Board
Murfreesboro Water & Sewer Department
Nashville Metro Water Services

Texas
Austin, City of
Dallas Water Utilities
Denton, City of
El Paso Water Utilities
Fort Worth, City of
Houston, City of
San Antonio Water System
Trinity River Authority

Utah
Salt Lake City Corporation

Virginia
Alexandria Sanitation Authority
Arlington, County of

Fairfax County
Hampton Roads Sanitation District
Hanover, County of
Henrico, County of
Hopewell Regional Wastewater Treatment Facility
Loudoun County Sanitation Authority
Lynchburg Regional WWTP
Prince William County Service Authority
Richmond, City of
Rivanna Water & Sewer Authority

Washington
Everett, City of
King County Department of Natural Resources
Seattle Public Utilities
Sunnyside, Port of
Yakima, City of

Wisconsin
Green Bay Metro Sewerage District
Kenosha Water Utility
Madison Metropolitan Sewerage District
Milwaukee Metropolitan Sewerage District
Racine, City of
Sheboygan Regional Wastewater Treatment
Wausau Water Works

Australia
ACTEW (Ecowise)
South Australian Water Corporation
Sydney Water Corporation
Water Corporation of Western Australia

Canada
Lethbridge, City of
Regina, City of, Saskatchewan
Toronto, City of, Ontario
Winnipeg, City of, Manitoba

New Zealand
Watercare Services Limited

United Kingdom
United Utilities North West (UUNW)

STORMWATER UTILITY

California
Fresno Metropolitan Flood Control District
Los Angeles, City of, Department of Public Works
Monterey, City of
Sacramento, County of
San Francisco, City & County of
Santa Rosa, City of
Sunnyvale, City of

Colorado
Aurora, City of
Boulder, City of

Georgia
Griffin, City of

Iowa
Cedar Rapids Wastewater Facility
Des Moines, City of

Kansas
Overland Park, City of

Kentucky
Louisville & Jefferson County Metropolitan Sewer District

Maine
Portland Water District

North Carolina
Charlotte, City of, Stormwater Services

Pennsylvania
Philadelphia, City of

Tennessee
Chattanooga Stormwater Management

Texas
Harris County Flood Control District, Texas

Washington
Bellevue Utilities Department
Seattle Public Utilities

STATE

Arkansas Department of Environmental Quality
Connecticut Department of Environmental Protection
Kansas Department of Health & Environment
Kentucky Department of Environmental Protection
New England Interstate Water Pollution Control Commission (NEIWPCC)
Ohio River Valley Sanitation Commission
Urban Drainage & Flood Control District, CO

CORPORATE

ADS Environmental Services
Advanced Data Mining International
Alan Plummer & Associates
Alpine Technology Inc.
Aqua-Aerobic Systems Inc.
Aquateam–Norwegian Water Technology Centre A/S
ARCADIS
Associated Engineering
Black & Veatch
Blue Water Technologies, Inc.
Boyle Engineering Corporation
Brown & Caldwell
Burgess & Niple, Ltd.
Burns & McDonnell
CABE Associates Inc.
The Cadmus Group
Camp Dresser & McKee Inc.
Carollo Engineers Inc.
Carpenter Environmental Associates Inc.
CDS Technologies Inc.
CET Engineering Services
Chemtrac Systems Inc.
CH2M HILL
CRA Infrastructure & Engineering
CONTECH Stormwater Solutions
D&B/Guarino Engineers, LLC
Damon S. Williams Associates, LLC
Earth Tech Inc.
Ecovation
EMA Inc.
Environmental Operating Solutions, Inc.
Environ/The ADVENT Group, Inc.
Fay, Spofford, & Thorndike Inc.
Freese & Nichols Inc.
ftn Associates Inc.
Gannett Fleming Inc.
Garden & Associates, Ltd.
Geosyntec Consultants
GHD
Golder Associates Ltd.
Greeley and Hansen LLC
Hazen & Sawyer, P.C.
HDR Engineering Inc.
HNTB Corporation
Hydromantis Inc.
HydroQual Inc.
Infilco Degremont Inc.
Jacobson Satchell Consultants, Inc.
Jacques Whitford NAWE, Inc.
Jason Consultants LLC Inc.
Jordan, Jones, & Goulding Inc.
KCI Technologies Inc.
Kelly & Weaver, P.C.
Kennedy/Jenks Consultants
KMK Consultants
Komline Sanderson Engineering Corporation
Larry Walker Associates
Limno-Tech Inc.
The Low Impact Development Center Inc.
Malcolm Pirnie Inc.
Material Matters
McKim & Creed
Metcalf & Eddy Inc.
Monteco Corporation
MPR Engineering Corporation, Inc.
MWH
NewFields Water Resources, LLC
O'Brien & Gere Engineers Inc.
Odor & Corrosion Technology Consultants Inc.
Original Engineering Consultants, Ltd.
Oscar Larson & Associates
Parametrix Inc.
Parsons
Post, Buckley, Schuh & Jernigan
Praxair, Inc.
RMC Water & Environment
Ross & Associates Ltd.
Rothberg, Tamburini & Windsor, Inc.
SAIC
Savin Engineers
Siemens Water Technologies
Stantec Consulting Inc.
Stearns & Wheler, LLC
Stone Environmental Inc.
Stratus Consulting Inc.
Synagro Technologies Inc.
Tetra Tech Inc.
Trojan Technologies Inc.
Trussell Technologies, Inc.
URS Corporation
Wade-Trim Inc.
Westin Engineering Inc.
Weston Solutions Inc.
Woodard & Curran
Zenon Environmental Inc./GE Water Process Technology
Zoeller Pump Company

INDUSTRY

American Electric Power
American Water
ChevronTexaco Energy Research & Technology Company
The Coca-Cola Company
Dow Chemical Company
DuPont Company
Eastman Chemical Company
Eli Lilly & Company
Merck & Company Inc.
Premier Chemicals LLC
Procter & Gamble Company
Thames Water Utilities
Severn Trent Services Inc.
Suez Environnment
United Water Services LLC

Note: List as of 7/10/07

WERF Board of Directors

WERF Research Council

IWA Publishing's authorised EU representative for General Product Safety Regulations is Diane D'Arras, 15 rue Duret, 75116 Paris, France, e-mail: safety@iwap.co.uk.

Printed and bound by CPI Group (UK) Ltd, Croydon, CR0 4YY
12/05/2026
02108879-0010